40 Maps That Will Change How You See The World

Quarto

First published in 2024 by Ivy Press,
an imprint of The Quarto Group.
One Triptych Place
London, SE1 9SH,
United Kingdom
T (0)20 7700 6700
www.Quarto.com

Design Copyright © 2024 Quarto
Text Copyright © 2024 Alastair Bonnett

A catalogue record for this book is
available from the British Library.

ISBN 978-0-7112-9353-3
Ebook ISBN 978-0-7112-9355-7

10 9 8 7 6 5 4 3 2 1

Design by Intercity

Publisher: Richard Green
Editor: Katerina Menhennet
Senior Designer: Renata Latipova
Production Controller: Rohana Yusof

Printed in Malaysia

40 Maps
That Will Change How You See The World

Alastair Bonnett

IVY PRESS

Contents

Introduction

I'm making you a promise. These forty maps will change the way you see the world. How do I know? Because that's what they did for me. My method for choosing them was simple. As soon as I discovered a map that gave me an electric jolt – made me jump up and want to grab someone, anyone, and tell them about it – then I knew. It's true that my tumbled speechifying always failed me. I had to show not tell. Mere words can never do justice to a great map.

You have to see a map to know it. You have to see the fragile nest-like beauty of a Polynesian stick map (see map 8, page 40); the pastel colours and extraordinary detail of the latest maps of Mars (see map 37, page 166); the saucy connotations buried in the map of love (see map 24, page 106); or get your head round mapping in three dimensions, because that's the only way we can navigate vertical cities such as Tokyo (see map 18, page 82) ... these maps disorient and reorient, they force us to reimagine things we thought we knew or took for granted.

Many of the forty maps I've collected show us perspectives profoundly different from our own. Aztec maps (see map 6, page 30), for example, are fantastical but also compelling, as they put us in direct, immediate, touch with cultures smashed up and shunted aside by the juggernauts of colonialism and globalization. Our 'modern' maps are all roads and cities but many 'pre-modern' maps had room for magic and storytelling. Those quote marks I hang on the word 'modern' mean something. This is not a book for those who do not like their assumptions tested or for Eurocentrics who think the West invented everything. Take a look at the ancient Chinese grid-square (see map 2, page 12). It's as rational, logical and modern as most twenty-first century

maps, but this one was created in the twelfth century.

There is no universal timeline in map-making, no simple story of an ascent from the simple to the sophisticated. Take another look, this time at our first map (see page 8), found on a wall excavated in Turkey. It was painted 4,000 years *before* the stones of the first pyramid were dragged into place, which makes it 9,000 years ago. Maps like this plunge us into the deep past and are truly exotic but they also recalibrate history, messing with neat notions of old and new, modern and premodern, us and them.

Maps can take us to other worlds. One of brightest horizons for cartography is outer space and I've included maps that represent the cutting edge of 'astrogeography'. They are created by scientists, and just like the map of the Earth's gravity (see map 30, page 134) or the nexus of trees and fungi (see map 28, page 128), they are things of beauty. Mapping is fast becoming a key tool for science, yet maps are art as well as science. They are both but cannot be reduced to either: maps work on their own terms and have their own traditions and ways of seeing.

Humans have been making maps for tens of thousands of years but today we are entering a new phase in map-making. Everyone with a phone is part of it: we are all constantly locating ourselves and being

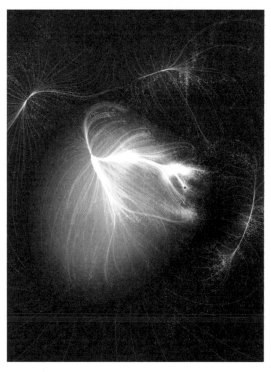

The Laniakea Supercluster

located. Maps are everywhere: communicating and organizing what we do and who we are. They measure the pulse of a plugged-in society.

The twenty-first century is a golden age of mapping – a bold claim that is demonstrated across these pages – but this only makes it more important that we open our eyes to the many routes and roots of mapping. Not just our minds. Our souls. When you look upon a Buddhist map from Burma (see map 9, page 44), which offers us a world shaped as a teardrop, shattering into islands, with a single green tree standing at its apex, you're being invited on a journey to another shore.

This book has been a lot of fun to write but it's also an act of tribute. It's a tribute to map-makers from distant times whose names are now long forgotten but also to today's lively generation of map-makers. Apart from the few we drew ourselves (see maps 20 and 31, pages 90 and 138), the maps in this book are originals, and we thank the many talented map-makers who have given us permission to use their work. Maps are one of the most powerful instruments of communication and discovery for twenty-first century researchers, in large part because with new software, computing power and satellite technology, it is possible to fashion maps of startling levels of complexity and accuracy. We are coming to understand that maps are not just illustrations or useful guides, they are tools of discovery that illuminate the innermost workings of society and nature.

There is a broad arc to this book. It goes from the very ancient to the very latest, and from planet Earth to outer space. But readers can dip in anywhere. Every page has its own story to tell.

Each of these forty maps is a disruptor. These are maps that challenge our world view, sometimes giving it a nudge, quite often knocking it off the table. Yet this is also a book of pleasures. The maps I've gathered are evocative, sensuous, exciting. They tell many tales and have much vital information to impart, but I hope you agree with me that each, in its own unique way, is an object of delight.

Map 1
Wall Painting, Çatalhöyük, 6000 BCE

A 9,000-year-old town map

Four thousand years before Egypt's first pyramid, there was a bustling town in southern Turkey. Its industrious inhabitants built and decorated neat, square mudbrick houses and buried their dead close by. At its peak about seven thousand people lived in what we call Çatalhöyük, an archaeological site that lies some 140 kilometres (87 miles) south of the twin-coned volcano of Mount Hasan.

This map is a long wall mural and was found inside one of the excavated rooms. On the left side you can just about see the volcano, picked out with spots, and its two peaks. Below it there is the town. It's a dense jumble of geometric shapes that look a lot like the houses that are so characteristic of Çatalhöyük.

It's a compelling image: the first map of the first town; a pre-historic fingerpost to an urban future.

Despite the site being so old we can tell some things about the people who lived here. We know their economy was based on farming and herding cattle and sheep and that their town was a maze, with the houses crammed together. Less than ten per cent of the site has been uncovered, so new discoveries are always round the corner. Recently the first street was found. Before then it was thought that Çatalhöyük had no streets and the only way of getting about was by clambering across the roofs. The flat roof tops were important. The houses don't have doors but openings in the ceiling. Ladders were used to get outside. The roofs functioned as plazas and communal areas.

This was a tidy place: the Çatalhöyükites were not slobs. Hardly any rubbish has been found in the houses and there were allocated spaces for debris and sewage. Another thing we know is that the people who lived here had a close relationship with the dead. They were buried amid the living – bodies have been found beneath floors, under beds and hearths – and their remains were passed round. Skeletal remains, many painted, appear to have been shared between households. We have no idea what their belief systems were, if they had a religion or not, but we can guess that the dead had an important role, a real presence, in ritual and daily life.

Çatalhöyük challenges us. It is nine thousand years old, but it isn't just dust and stone: it poses questions. Some of the most intriguing ones come from the fact that it doesn't appear to have had social hierarchy. There is no palace, no big houses for the rich, no signs of kings and queens. The houses are all similar. This was a communal society in which labour and food were shared. There are no clear signs of a pecking order. This does not only apply to class. From what we can tell gender discrimination did not exist in Çatalhöyük. The most famous artefacts found here are stately female statuettes, whose ample proportions suggest they had something to do with fertility. The Seated Woman of Çatalhöyük is the best known. This majestic and unapologetically fat female form rests her hands on two lionesses. She was found in a grain

Seated Woman of Çatalhöyük

bin, a clue that indicates that her presence was thought to bring good luck to the harvest.

Far more female figures have been found than male ones. Some scholars have concluded that Çatalhöyük was a matriarchy. However, an even more radical explanation appears to be case. Evidence from skeletons shows that men and women received the same amount of food and that they did the same work, spending the same amount of time outside. They lived side by side and had the same funeral rites. The simple yet profound conclusion appears to be that the two sexes were equal in Çatalhöyük.

Let's return to the map. More specifically, to those who question it, who say it's not a map but something else. Look at those 'house' shapes. They are, these voices tell us, more likely to be representations of an animal skin, such as a leopard, than an attempt to draw any kind of plan. Yet another theory is that this is a map but nothing like maps as we know them: it does not attempt accuracy or direct semblance but symbolism, pointing to long-forgotten traditions of earth magic and geomancy.

These theories have their merits. But aren't we forgetting that images can have more than one meaning? Isn't it likely that our map was understood and used in more than one way? That its meaning shifted across decades and generations?

James Mellaart, the English archaeologist who first uncovered Çatalhöyük and this map, was struck by how much 'the variations and irregularities in the drawing' resembled the town. He also theorized that the mountains were represented because of their importance as a source of 'obsidian, the volcanic glass from which they made their tools and weapons, beads and mirrors'. On this point he was wrong. The inhabitants did use obsidian but chemical analysis shows it came from much further away, 380 kilometres (236 miles) away in fact.

Mellaart died in 2012, his credibility in question. He had a bad habit of making up stuff that didn't fit with his theories and this sullied his reputation. However, Mellaart did discover Çatalhöyük, he knew it, and, looking at those shapes, those mountains, his conclusion that this mural 'is a representation of a neolithic town', remains not just plausible but the best explanation of what we are looking at.

Excavations at Çatalhöyük are ongoing and each year sees fresh discoveries. Recently a new neighbourhood has been unearthed. Future digs will surely provide new insight into the meaning of the mural, perhaps they may reveal more maps. Ali Umut Türkcan, head of the site excavation, emphasizes that just a small fraction of the site has been uncovered. This multi-layered and important place, represents, as Türkcan tells us, 'the beginning of urban culture'.

Its inhabitants are *very* distant from us. Nine thousand years is longer than any of us can imagine. Yet in their clean houses, their love of decoration, their fascination with the dead, and their sense of order, we know them and we recognize ourselves.

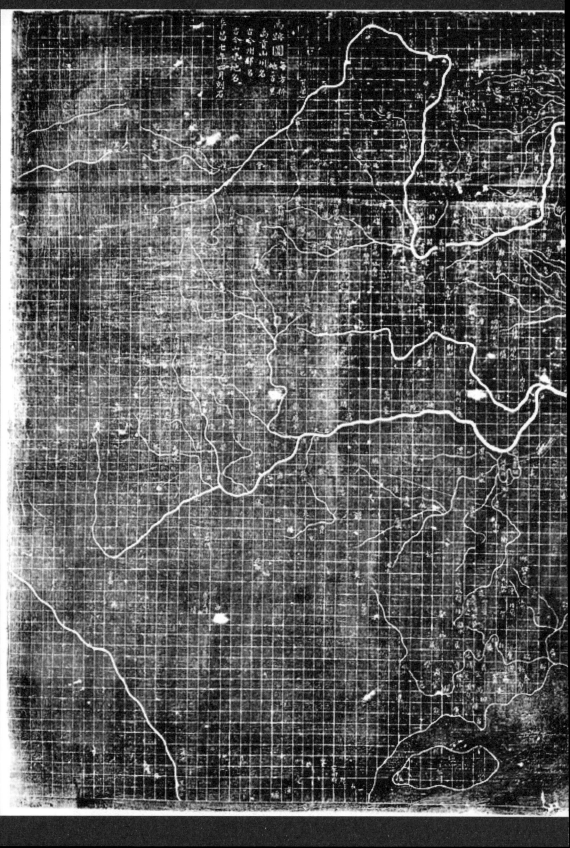

Map 2
Map of the Tracks of Yu, 1137

The first modern map

For anyone who thinks that modern life was invented in the West and shipped about the planet to 'less-sophisticated' folk, this map will come as a surprise. It covers thousands of kilometres of east and south China and was carved in stone in 1137. It's not a hundred per cent accurate. The north-eastern parts of the country are distorted, to allow all the key rivers to be included. Yet the intent is recognizably scientific.

The land is divided into a grid square (something European map-makers only started doing 700 years later) and towns, coastlines and rivers are placed, not by virtue of tradition or educated guesswork but according to a rational methodology. This effort, this desire for logical order, makes this map totally modern.

Our map does not look like a lump of stone and that is because it isn't: it's a rubbing, a copy. This is how the map was designed to look. The stone block that contains the original, chiselled out by unknown hands, isn't a monument but a printing block. The point of carving the map was to allow numerous rubbings to be made. Copies were sent far and wide. The original stone is one of many ancient 'information boards' you can see in the city of Xi'an's Forest of Stone Steles Museum.

What we have here is not a just a map whose technical sophistication is astonishing, but a printing process that points to a complex and co-ordinated system of knowledge dissemination. While all this was going on, European maps were more fairytale than science; they showed you where Jerusalem might be (in the middle) and where various monsters dwelt but as guides to the shape and placing of countries, rivers, cities and coastlines, they were useless.

China used modern mapping many centuries before Europe. And this map covers a huge area. Those thick white lines are the Yellow and Yangtze Rivers and we can also see the island of Hainan, 'China's Hawaii', at the bottom, sitting in the South China Sea. The whole map is covered in place names and pinpoints about four hundred administrative districts and seventy mountains. There are also five lakes, the white blobs, which are so accurately located so we can name them; they include Taihu Lake, Dongting Lake and Poyang Lake.

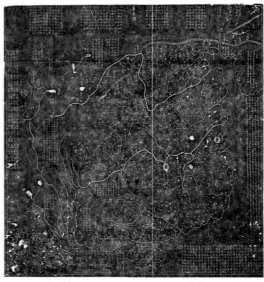

This is not a map of roads but of rivers, of which there are eighty. White waterways snake across its surface and help us understand that ancient Chinese civilization was water-borne and centred itself along valleys.

Looking more closely we might make out some text at the top of the map. It hints that this impressively rational map is also a celebration of a legendary man. It reads: *Map of the Tracks of Yu* and tells us it was created in honour of Yu the Great.

A rubbing of the 'Map of China and the Barbarian Countries' from the Library of Congress

不伐不矜　振古莫及　惠洒時言　九功由立　盧敬在躬　殷中九訊　克勤于邦　烝民乃粒　禹

Yu the Great, by Ma Lin
(circa 1180-1256),
hanging silk scroll

Who was Yu the Great? It's a fascinating story. Yu the Great pre-dates written records. He appears to have lived sometime between 2000 and 2200 BCE and he was also called 'Yu that controls the flood', for he is credited with channelling China's rivers. Oral histories have Yu speaking of a great flood. 'The inundating waters seemed to assail the heavens', Yu tells us, 'so that the people were bewildered and overwhelmed'. He leapt into action, diverting and channelling the water and dredging rivers. He proclaimed that he had 'opened passages for the streams throughout the nine provinces and conducted them to the seas'.

Yu is said to have had the common touch. He ate and slept with his workers and got his hands dirty, working for thirteen years on his great task. It is also said though, that Yu had some extra non-human help: a channel-digging dragon and a giant mud-hauling tortoise were allegedly at his command.

Yu's efforts have been much praised and embellished over the years. When this map was carved out, he was a folkloric and unifying historical figure. Even today he's a captivating, unusual, hero. The idea that civil engineers could be the stuff of legend wasn't something typically seen in Europe until the late nineteenth century. It seems that the Chinese have long valued innovation and useful labour and had a clear sense that progress – the betterment of the nation – was both needed and possible.

What isn't on a map can be just as revealing as what is. There are plenty of grid squares, lakes and rivers, but the *Map of the Tracks of Yu* does not have any political borders. We know that borders, and lots of them, did exist, in 1137. At that time China was a thousand-piece jigsaw of competing kingdoms. The city of Xi'an, where the map was created, was its own tiny kingdom, ruled by a local war lord. So what is going on?

It could be argued that the map is a wishful one: an assertion of the essential unity of all the lands where a Chinese, Confucian culture held sway. Read this way, it's as much a cultural as a practical statement: it proclaims that, at some deep level, the Chinese are one people. There is another map, of nearly the same date, on the other side of the stone, that also speaks of a powerful sense of national identity. Its title is *Map of China and the Barbarian Countries*. It shows the Great Wall and gives potted descriptions of Korea, Japan and other 'barbarian' places.

This map has long been a favourite of map historians. Joseph Needham, a British historian of Chinese science, spent his life proving how many aspects of the modern world had taken root in China long before they eventually arrived in the West. Needham was right about many things, and he was right about this map when he wrote that the *Map of the Tracks of Yu* was 'the most remarkable cartographic work of its age in any culture'.

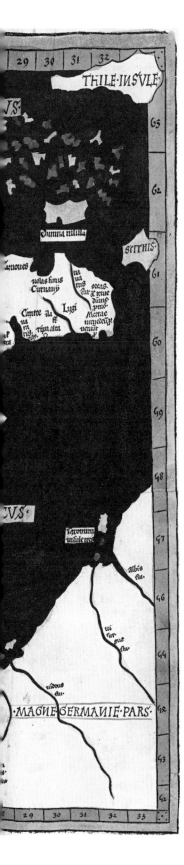

Map 3
Ptolemy's 'Prima Europe tabula', published in Ulm, Germany, 1486

The oldest map of the British Isles

This map is around 1,800 years old. Or, at least, the original was, which was created in the second century CE by the Roman geographer Ptolemy. This copy, made in 1486, is one of the earliest surviving maps of the British Isles. The yawning centuries between Ptolemy and his fifteenth-century imitators raise an intriguing question. Can you imagine taking a map created over a thousand years ago and thinking, 'Wow, this will be useful!'?

Reverence for the ancients was already starting to crumble when this copy was created. Respect for the glory that was Rome was being mixed with the idea that useful knowledge comes from experience, of going out into the world and discovering things for yourself. Over the following decades, in the wake of Christopher Columbus's voyages to the Americas, the demand for maps that actually got things right and helped people – especially navigators – get from A to B, became unstoppable and archaic renditions of geography began to seem as quaint as they do today.

None of that takes away from the achievement of Ptolemy. This is just one of his many maps. A version of his map of the known world, *Ptolemy Cosmographia* by Nicolaus Germanus, 1467, is also reproduced overleaf, another copy from the fifteenth century. There are good reasons why Roman geography impressed later generations. Despite the absence of any of the techniques and technologies that modern cartographers rely on, the countries and coastlines in Ptolemy's maps remain instantly recognizable.

Ptolemy's accomplishment was considerable, but can we say the same for the creators of these reproductions so many centuries later? They are beautiful objects, yet they speak of the inertia of pre-modern Europe, of a long period when culture and science were not actively pursued but burnished and passed down, handed on like treasure.

The first thing to notice about the map is that the top part of Britain goes off at a right angle. Scotland is bent radically to the east. It's a glaring error, which would have been obvious to anyone who knew the island. Ptolemy seems to have passed on this mistake from an earlier source which was equally off-kilter. Putting this bit of weirdness aside, Britain and Ireland are familiar. The island of Hibernia (Ireland) on the left-hand side of our map is sort of the right size and sort of the right distance from Britain on the right. The peninsula of England's south-west is there and so too the bulge of Wales to the west just below and to the right of Ireland.

This 1486 version has lost some of Ptolemy's place names and there aren't many that ring modern – or fifteenth-century – bells. We might make out 'londinu', in the centre just above the part of Britain which juts out to the bottom left of our map which I guess is London, and the island of Britain is called 'Albion Insula', the island of Albion.

Some parts are more familiar than they look. Adrift off Ireland's south-east shore and coloured in yellow is an island labelled Mona. This is Anglesey, which was called Mona by the Romans. The Welsh adopted this name and, to this day, Anglesey in Welsh is 'Ynys Mon' or Mona Island.

Another seemingly misplaced island can be found in the eastern corner of England, where we find an island shaded in red called 'Counus'. There is some mystery to this. Could it be Canvey Island? The Elizabethan geographer William Camden thought so and claimed that 'Ptolemee maketh mention' of it. Perhaps, but probably not. Canvey Island is not, in fact, an island, it's a coastal bit of Kent and this is a part of Britain the Romans knew well. So what is Counus? Islands do come and go in the English Channel. It may be what we're seeing is the trace of a now-disappeared island.

Not all the islands are in the wrong place. Another, off England's south and also coloured in red, is recognizably the Isle of Wight. Here it is called 'Occes', which may be the name of a tribe that lived there.

Many of the settlements located are Roman camps. Again, that needling question: why would fifteenth-century map-makers go to such lengths to record such archaic knowledge? It would not have been hard to find out what the contemporary positions and contemporary names were of the towns of Britain and Ireland. People in the fifteenth century knew that the legendary island of Thule, which the Romans romantically thought was at the edge of the world, the last of places, was just that, a legend. But there it is, top right. Uncritical respect for the past – for the classics – leads to a backward-looking geography.

Turning to Hibernia (Ireland) there are just as many puzzles. There are plenty of place names but how they relate to modern towns is far from clear. One possible lead is 'Eblana', which we can see on the east coast, half way down, just about where Dublin is today. Dublin's claim to have a two-thousand year history is based on this map. Some historians have poured icy water on the Eblana–Dublin theory. There are no remains that point to people living in Dublin before the Vikings. But is that such a clincher? None of the settlements mentioned on Ptolemy's map was a big place; they were villages and camps, and they were haphazardly located. Aside from Roman walls and villas remains are few. Let's not forget that Ireland was never conquered by the Romans. Eblana could be Dublin.

Even more intriguing, what is that green oval in the middle of Scotland? It's clearly a great forest, and here it is labelled Caledon. It has been speculated that it was placed here by Ptolemy as an excuse for the failure of the Romans to conquer the top end of Britain. However, in another example of the influence of this map, the idea of a lost Scottish 'great wood' persists to the present. In his book *The Great Wood* Jim Crumley explores the legend. He concludes that 'its reputation' was created by the Romans to evidence the wildness of these northern reaches. It's a likely explanation. This is a map built on travellers' tales, hearsay and speculation. It's also a spring, a font for many stories, real and otherwise, that ripple down the centuries.

Ptolemy's world map, Nicolaus Germanus, 1467

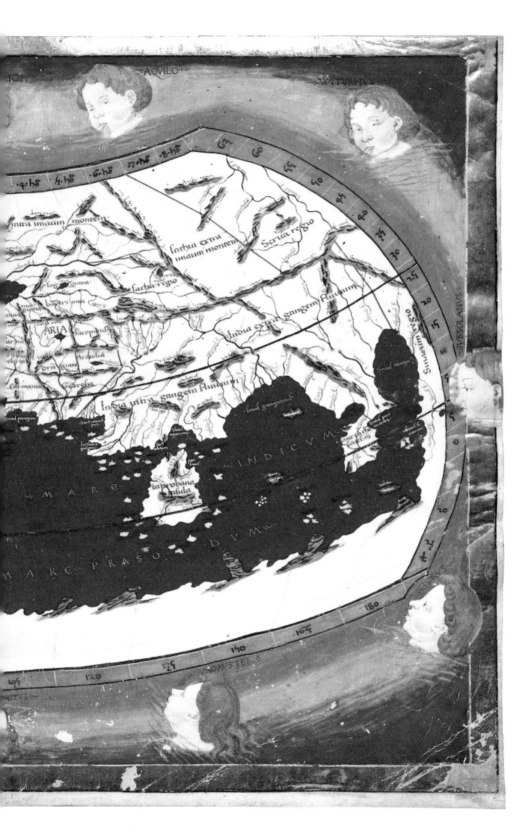

Map 4
'A World Chart Showing the Barbarians', 1418(?)

When China discovered the world?

History is flipped on its head. More than seventy years before Columbus stepped on the sandy shores of the Bahamas, the Chinese had mapped not only the Americas but pretty much the whole world. It's an extraordinary story.

Inked on bamboo parchment, what has become known simply as the '1418 map' was found by chance in 2001. A Chinese amateur historian called Liu Gang was sifting through the stock of a local antique dealer in Shanghai. He caught sight of this old brown scroll, knew it was special and paid US$500 for it.

Its big international release came in 2016. Unveiled in Beijing and in London, and splashed in newspapers the world over, the map was a sensation. *The Economist* proclaimed a global game-changer: 'the world and all its continents were discovered by a Chinese admiral'. They also got in a dig: 'The Chinese, having discovered the extent of the world, did not exploit it, politically or commercially.'

Let's take a closer look at it. A caption explains that it's a copy of an original, and that it was 'drawn by Mo Yi Tong'. It reads, 'in the year of Qianlong Gui Wei [1763] by imitating a world chart made in Ming Yongle [1418] showing the Barbarians' and declares itself a 'general chart of the integrated world'.

It's got everything. The shapes of Asia, Africa and Europe are instantly recognizable. North America has been inflated as if with a bicycle pump but both the north and south of the continent are roughly where they should be. To draw all these coastlines, they would have had to have been navigated. Even more mind boggling, there is Australia, the large island in the middle to the right and Antarctica peeking out at the bottom.

What else can we see? There is a lot of text. On the southern end of Africa is written 'The skin of people here is like black lacquer. Their teeth are white, their lips are red, and their hairs are curled'. For North America: 'There are more than one thousand tribes and kings here'. On South America: 'Human beings are used as sacrificial victims, and people pay obeisance to fire'. China gets a somewhat different treatment: 'The most primary country under heaven'.

The judgements are contentious, but they are numerous and taken together they represent a compendious global survey.

Perhaps the most impressive thing about the 1418 map is its pedigree. It depicts the discoveries of the legendary Chinese explorer, Zheng He. Zheng He was born in 1371 and statues of him, bold, martial and stocky, are popping up in many Chinese cities. They are all recent, for Zheng He is a key symbol of China's newfound image of itself as an outward looking and muscular global player. He is often just called 'the eunuch' for when Zheng He was eleven his home province was invaded and he was captured, castrated and put to work in the imperial court. His traumatic boyhood didn't tether his ambition. Zheng's quick wits were recognized, and he was promoted rapidly, becoming a diplomat, mariner and, eventually, the admiral of successive fleets.

Zheng He's 'treasure ships' roamed the oceans between 1405 and 1435 and they were huge. His first voyage had 317 ships and almost 28,000 crewmen. Even before the 1418 map was unveiled it was known that Zheng He had voyaged thousands of miles from home. He sailed all over south-east Asia and on to India, Arabia and the Horn of Africa. These were, indeed, voyages of trade not conquest. In return for gold, silver, porcelain and silk, Zheng He sent luxuries and novelties back to the imperial court, ivory but also ostriches, zebras, camels, and giraffes.

Zheng He's fleet was called the Star Raft and his exploits have long been associated with map-making. Zheng's charts covering Asia and the east coast of Africa are detailed and compelling. The 1418 map did not spring from the blue. It is a radical leap, but it is an extension of an existing, verified history and the exploits of one of the world's greatest sea captains.

It's also a map people want to see. For too long, the history of the world has been told as a European story. Today this approach doesn't wash. What more compelling evidence than this! The 1418 map has been seized upon and delighted in.

Nevertheless, it is still the subject of argument and controversy. I'd rather not sit on a rickety fence. The 1418 map is a fake. Let us take another, more sceptical, look at it. First it is worth knowing that none of the other Chinese maps of the period look anything like it and there are no Chinese records of the extensive worldwide journeys that would have been required to complete it.

It is also odd that its rendition of China (centre-left), which we know had been expertly mapped by the Chinese, is so cursory, as if dashed-off. It gets worse. One text box tells us that the Himalayas are the highest mountains in the world. So not only were all the world's coasts but all the world's mountains surveyed and measured? In 1418?

Professional map historians are scathing about the 1418 map. Geoff Wade, a cartographer from the National University of Singapore, bluntly states that there 'is absolutely no possibility that it is anything but a product of the last fifty years, and quite possibly of the last five years'.

The 1418 map was conjured into global attention because so many people want it to be true. The historical reality it points to is that China is back and making waves and all sorts of proof of its ascendance and pre-eminence are being found. Paradoxically, the 1418 map obscures the fact that China is home to the most impressive ancient map-making tradition in the world and that Zheng He's discoveries were indeed world changing. There is still widespread ignorance of these facts; an ignorance that plays out into a queasy sense that exaggeration and making Chinese history mimic European history is necessary to rebalance the historical record. China is being understood as Europe 2.0. It isn't: it's history is not lesser, nor is it greater, but it is different.

A tribute giraffe sent to China,
16th century, hanging scroll

أرض حديد

Map 5
World Map from the *Tarih–i Hind–i Garbi*, 1580s

The world on its head. The Ottoman
Empire eyes the New World

Around the coasts of America, Africa and Asia the wind fills the white sails of explorers' ships. It's a recognizable image but an odd one, because it is upside down. Flip the map on its head and it looks like something we're familiar with. It's pretty accurate around the Mediterranean and Middle East – which gives us a clue to its origin – but North America is misshaped and there is a weird land mass stretched across the Atlantic, in the middle at the bottom of our map.

At the top, the map shows what looks like Antarctica; most likely a representation of the yet undiscovered, fabled southern continent later located and named Australia.

What's going on? And why is it the 'wrong' way up? The makers of this map might ask the same question of maps which we imagine are the 'right' way up. The world has poles but either one could be the top or bottom, and either one could be called 'North' and 'South'. The orientation of the planet seen on this map, drawn in the 1580s in Constantinople (a city later rebranded as Istanbul), is not actually wrong, just different.

It was created at the height of the Ottoman Empire. It was a vast realm: Constantinople ruled over Turkey, much of north and east Africa, and substantial parts of Asia, Europe and the Middle East. But this isn't a map looking back at past glories. Its agenda is plain from the book it comes from, a book that describes the riches to be had in the 'new world', the Americas (the *Tarih-i Hind-i Garbi or* 'History of the West Indies').

The Ottoman Empire was huge but at least some voices in its ruling circles wanted more. Information was reaching the court about the Americas, the 'new world', which must have been both intriguing and galling. Here was a land ripe for conquest, rich and defenceless, yet being gobbled up by empire builders from Christian countries. This isn't just a map but a cry of alarm and a statement of intent: to have a share in the gold and land that the Spanish and Portuguese were grabbing for themselves.

Geographers from predominantly Muslim countries had a long tradition of drawing the world map with what Westerners think of as the south at the top. It was a cartographic tradition established over hundreds of years. Many of the earlier versions of these 'upside down' maps placed Mecca as the centre of the world. By the same logic that made Europeans think that an orientation that placed Jerusalem and, later, Europe centre stage was sensible, maps drawn by Ottoman, Arab and Iranian geographers assumed Muslim countries were at the heart of the world.

This flipped map was following a well-worn formula. Yet the world was changing; the old formulas were failing. Even as their empire reached its zenith the Ottomans were confronted with the possibility that power was shifting, away from them and towards distant shores. The *History*

Bolivian silver mine from the Tarih-i Hind-i Garbi (c. 1600)

of the West Indies exhibits a fascination and curiosity about these exotic new lands. Its pages are full of illustrations and stories, many drawn from Italian sources, of strange marvels.

Among the many painted images that decorate the book we can find coconut, guava, cactus and banana trees, as well as tapirs, armadillos, bison, pelicans and jaguars. There are also more fanciful images, such as a tree fruiting the bodies of naked women. We also get portraits of the treasure that the Ottomans were missing out on, such as the Bolivian town of Potosí. Looming over the town is a green mountain, full of precious silver. You can see a black canal coming from the mountain, which was used to sluice out the ore. There are traders in the foreground. For an empire with many mines of its own, it's a tantalizing image of enormous but distant wealth. Elsewhere the unknown author salivates about a mine where 'gold nuggets worth ten dirhems were often found; sometimes it happened that one the size of an orange was found'.

Central to the authority of the Ottoman Empire was the claim that it was acting for and on behalf of Islam. Its rulers were not just sultans they were caliphs, successors to the prophet Muhammad. Just as the European conquerors of North and South America asserted their right and duty over it by virtue of their Christianity, the author of *History of the West Indies* makes much of the urgent need to spread Islam to these misguided regions. Writing of a Caribbean island that appears to be Hispaniola (today shared between Haiti and the Dominican Republic), the author explains that 'All worship the accursed devil and they make ugly idols in the forms of demons'. The writer then takes aim at the cruelty of the European invaders and pleads for Ottoman intervention.

In twenty years, the Spanish conquered all of that island and enslaved more than forty thousand; many hundreds of thousands fell prey to the sword. By the Lord God we always hope that the advantageous land will, in time, fall conquest to the heroes of Islam of exalted lineage and that it will be filled with the rites of Islam and be joined to the other Ottoman lands.

It didn't happen. The Americas did not 'fall conquest to the heroes of Islam'. Why didn't the sultan act? Perhaps he thought his empire large enough or that expansion east and south made more sense than getting into conflict with the rising powers of the west.

Yet this map is telling us that, for a moment, another history was being urged. It was a path not taken. If it had been, the world would look very different. Perhaps people in Brazil or the USA would be speaking Turkish. Perhaps we would have become familiar with different orientations, in many things, including what we imagine is the right way up for a map.

matlactli... /o riatoiani tlasoïuani guetsalecatzin
gmihquac llogu en. Halli ima cavalli ypiltzin
dotossanctos s uo mctancra ziatepe
guan stitzili

ollintzin

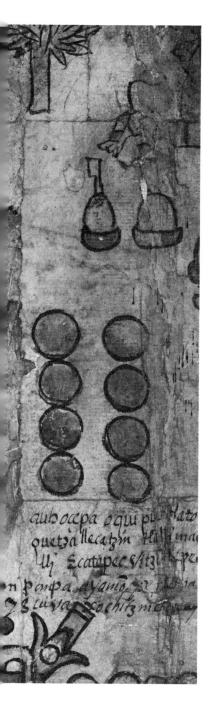

Map 6
Codex Quetzalecatzin (detail), 1593

Lord Quetzalecatzin's map: Aztecs versus colonial power

Lord Quetzalecatzin, dressed in a red cloak, sits on a jaguar fur, and behind him kneels his wife. It's a detail from a map that is all about land ownership. In the full version of the map, also reproduced on p33, we see Quetzalecatzin's male descendants stacked beneath him, along with their wives behind them, and it is these male heirs – more specifically the last and living ones – that used this map to stake a claim. 'This land is mine.'

We're in Mexico. The map shows part of the Puebla and Oaxaca regions. But we are looking at more than territory. This map also shows us time. This is a family tree, one that makes an assertion of land rights that reaches down from before the Spanish invasion, from the time of Quetzalecatzin in the fifteenth century, to 1593 when this map was drawn.

This is an Aztec map, full of Aztec imagery and symbolism and it makes use of the Nahuatl language, in hieroglyphic and other forms. Nahuatl was the *lingua franca* of the Aztecs, and it is still spoken by about 1.7 million people. The stylization of people, rivers, roads and pathways are all Aztec and the mixture of time (the circles and dots represent time periods) and space (the map shows a real landscape) is also typically Aztec.

Although the centuries have bleached the colours, we can tell this was once a brilliantly bright parchment. The colours were made from natural dyes; the blue, for example, made from indigo plant leaves and the red of Quetzalecatzin's cloak made from cochineal. Cochineal is a tiny cactus-dwelling insect, and when dried and crushed it yields a vivid colour. After the Spanish conquest, cochineal became a globally traded commodity, dyeing the red coats of British soldiers and the Catholic clergy's capes.

Yet this is not just an Aztec map. It is also a colonial one. It only exists and only survives because it was a response to a Spanish survey, an itemization of resources of who owned what. It's been named the *Codex Quetzalecatzin*, but it's not simply or purely an indigenous document and its land claim might be seen as more of a plea to new masters than an assertion of fact.

The label – *Codex Quetzalecatzin* – is a recent invention, disguising the fact that the name of the family shown was not 'Quetzalecatzin' but 'de Leon', an adopted Spanish name. The marginal notations tell us that the names of other elite families had gone the same way. There are the 'don Alonsos' and 'don Matheos'. It seems the Spanish honorific 'don' had become popular and that these families had been baptized, perhaps in one of the churches shown elsewhere on the map. As to the writing, it contains Aztec hieroglyphics but much of the Nahuatl is written in a Latin script and there is also plenty of Spanish.

So what do we have? Something complex. Something between and betwixt. Something important. The *Codex Quetzalecatzin* is one of the most important maps in the history of the Americas. It is important not because it is an indigenous map, or a colonial one. It is important because it shows transition and the birth of a *mestizo*, or 'mixed', society.

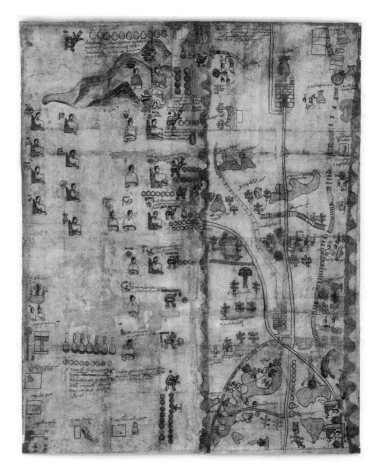

Codex Quetzalecatzin (full map), 1593

If we compare the *Codex Quetzalecatzin* to a fully pre-colonial map, one that features the same part of Mexico, the difference is stark. Nearly every pre–colonial map, along with untold numbers of other books, texts and artworks, were destroyed by the invaders and their followers. They were accused of being demonic, and burned. There are very few, very rare, survivals. The most significant, a map of the Apoala Valley in Oaxaca, the *Codex Nuttall*, is reproduced overleaf.

It does not look like a map. Today our idea of what maps look like is very literal. We think of them as tiny versions of reality as seen from above. Back then, back there, people had different ideas. They did have topographical maps, what we would think of as 'proper maps'. They are all lost but we know they existed because the conquistadors wrote about them. Hernán Cortes and Bernal Díaz del Castillo described indigenous itinerary maps and said they were used by travellers. The invaders also described maps that were used to show property ownership. Both of these traditions can be seen on the *Codex Quetzalecatzin* but there is another tradition that it barely hints at and which we see in this other image, which is contained in the *Codex Nuttall*, a super-rare pre-colonial survivor from the 1400s.

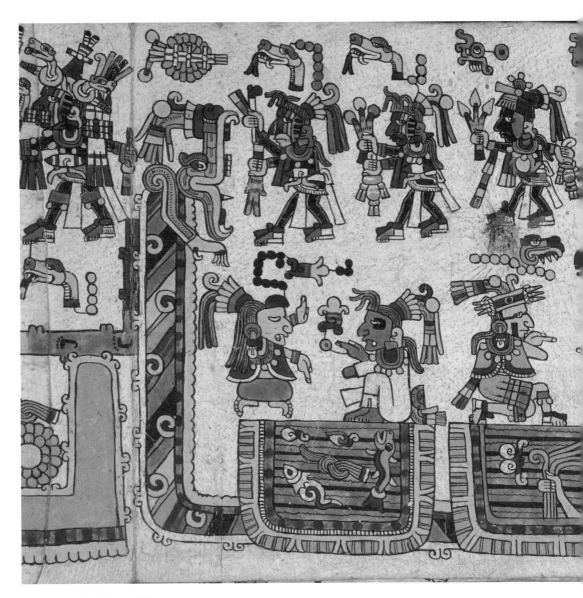

Codex Zouche-Nuttall, c. 1400s

34

The *Codex Nuttall* is a map of the Apoala Valley and it helps to know that the Apoala Valley is being shown in cross-section; the stripy shapes on the sides and bottom are the sides and bottom of the valley. Their stripiness indicates that this is soil and the double curls on their underside represent stone. There are two u-shaped forms on the floor of the valley; these are rivers. You may be able to spot fish and wavy, water lines. High up on the left side of the main valley there is an alarming monster with its mouth sprung open: this tells us there is a cave here. This Aztec map may look strange but, once we decode the symbols, things fall into place.

What is happening on the right-hand side of the valley? A naked half figure disappears into the cliff, on which stands a tree and there is also what looks like a waterfall. To understand it I turned to Professor Barbara Mundy, an authority on Aztec art. She tells us that there is a lot of birth imagery at work here, related to the origins and lineage of local people and that the half-body shows that this is the 'cliff of the childbirth' and that the tree 'may refer to the Apoala birth tree, from which important Mixtec [Aztec] lineages were born'. Professor Mundy brings us up to date by explaining that 'Apoala residents have recently identified' this very tree 'as a huge tejocote [Mexican hawthorn] or perhaps ceiba tree that once grew on the bank of the river above the waterfall'.

Aztec mapping traditions were intricate and complex but so many were thrown into the flames. Taken together, seen side by side, these maps show a moment of continuity: of traditions from the past bleeding into the present. They are both beautiful but I can't help thinking of all the maps we don't have, burnt and lost; atlases whose pages we'll never turn. The geographical imagination has grown and flourished across the centuries, it has countless roots and routes, but not a few have been cut down, pulled up and what we are left with are merely lovely fragments.

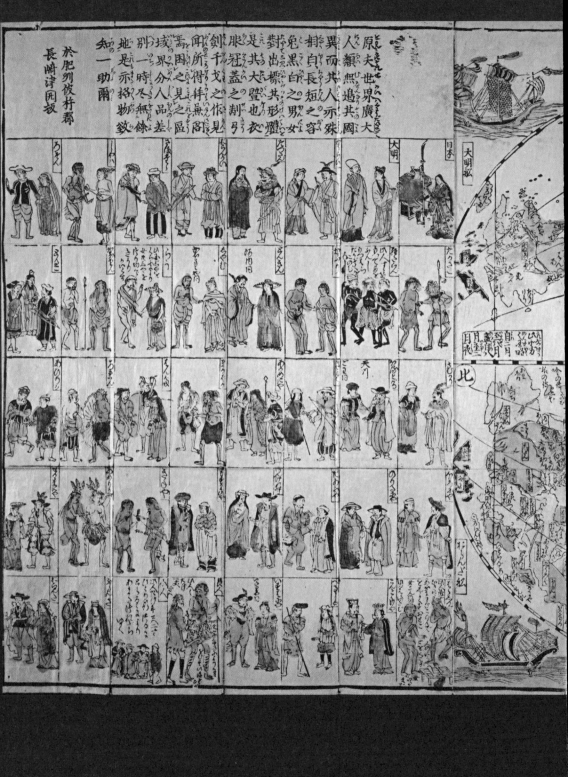

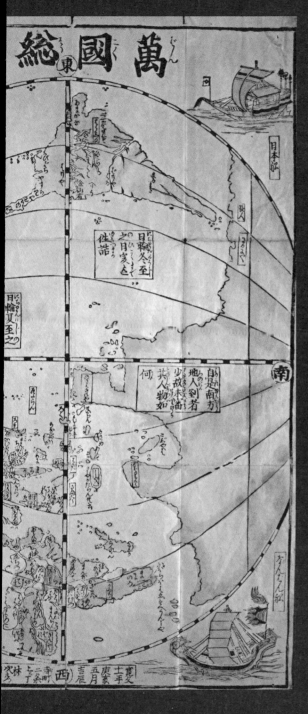

Map 7
Bankoku sōzu,
'Universal Map',
1671

All the world's people, as seen from Japan

People crowd this map. National and ethnic couples jostle, dressed in what the map's unknown creator thinks is their typical national clothing. There are forty types on display, some with potted descriptions, such as the man and woman from Brazil: 'These people do not live in houses; they like to live in caves. They eat human meat.' As an afterthought it is also explained that 'When a woman is about to give birth, the man's stomach hurts. Women feel no pain.'

The people who live on the 'Island of the Thieves' (which appears to be the Mariana Islands, which lie in the North Pacific) receive a similarly terse judgement. 'These people use ink to paint their body,' we are told and 'they only like to steal'. There are also some real exotics in the mix: dwarfs and giants.

Japan has much older maps, but this is the first truly global one. It took inspiration from Chinese and European models. In 1584 an Italian missionary based in China, Matteo Ricci, had created a *Map of the Myriad Countries* for Chinese readers. It had impact; it challenged China's ancient cartographic traditions and changed mapping in China and, later, in Japan too.

Yet this map is no faint copy. It's an original work, with opinions and perspectives of its own. The first thing we notice about it is that the world has an unusual orientation. The Americas lie at the 'northern' end and Asia in the middle and bottom. This makes Japan more or less central but quite what the thinking was that lay behind this orientation is not clear. When it was made, there was no fixed 'way up' for representing the world.

It's called the *Bankoku sōzu*, or universal map. The version we see here is a 1671 woodblock-printed copy of an earlier and larger original. It's an intricate, pleasing object, designed for display. This type of map would originally have been hung in the *tokonoma*, which is a space set off from a reception room in an affluent and traditional Japanese house. In the *tokonoma*, artistic and intriguing objects can be enjoyed and discussed.

In the seventeenth century, and for many years after, Japan had very limited contact with the outside world. Beyond its immediate neighbours, contact with foreigners was restricted to tiny numbers of Portuguese and Dutch. Most Japanese would have had not just no contact, but no knowledge, no notion of the many peoples shown on this map. It must have been an object of intense interest, and a centrepiece for amusement and debate.

'Dwarfs and the Giants', as depicted on Bankoku sōzu

'Cannibals', as depicted
on Bankoku sōzu

Perhaps these debates would have been muted, so as not to attract too much attention. The country's ruling elites were suspicious of outside influence. Multiple restrictions were in force against interaction with foreigners. The building of ocean-going ships was prohibited to stem the threat of contamination. The import of books almost ceased. This map may seem quaint to us, but understood in its context, this 'universal map' may have been seen as not just interesting but dangerous.

One nice touch is the four ships in each corner of the map. The upper right shows a Japanese ship, the upper left a Chinese ship; the lower right has a Portuguese trading ship and the lower left a Dutch ship. We can tell this from the script: there is little attempt to make any of the boats look non-Japanese.

The writing at the top of the ethnographic part of the map tells us what it is trying to achieve. This, it says, is a map about human diversity; and people come in all types and colours. 'The countries differ and people are also different,' it explains, adding that there are 'big and small ones, black and white men and women'. The author also points to differences in the 'making of clothes and headdresses, the production of bows, swords, shields and spears' and concludes that this map is 'an aid for understanding different things'.

The inclusion of dwarfs, giants and cannibals may seem a fantastical touch. Not so. It is in keeping with world maps created in Europe at this time and is likely to have been taken seriously. The note on the dwarfs tells is that they measure about 36 centimetres (14 inches) in height and to avoid 'being captured by cranes when going alone, they always walk together in groups'. This bit of folklore was likely taken from a Chinese encyclopaedia from 1609, which adds explanatory detail, mentioning that when 'birds encounter them, they swallow them. Therefore, when going out, they move in rows.'

The giants are isolated rather than preyed upon. They are said to live in the huge southern continent that fills up much of the right side of the map. 'From here,' it is explained, 'one encounters few people'.

The shape of Europe, which can be found bottom left of the world map, is haphazard but there are plenty of recognizable nations, including *Isuhaniya* (Spain) *Horutogaru* (Portugal), *Furansa* (France), *Itariya* (Italy), *Ingeresu* (England) and *Iheruniya* (Ireland). In other continents there are fewer named countries. The coloured regions refer to broad territories and ethnic groups. The western middle part of South America is one such region, where it is explained 'There are gold mines'. The north-eastern part of the Nort America is labelled *Amerika* but also *Nobafunsa*, or New France. Islands to the 'north' (in this map, west) of America and Eurasia are called 'Country of the Night'.

Japan opened up to the world in the late nineteenth century. It had spent centuries apart. What this map shows is that, even when isolated, people are still curious. Despite harsh restrictions, the desire to know the world, to extend the imagination out, to other peoples and other lands, is insatiable.

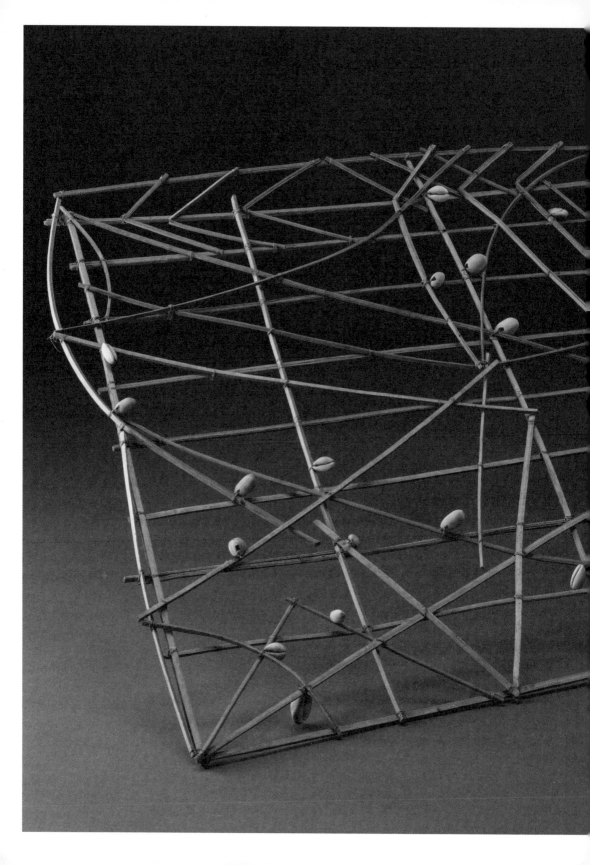

Map 8
Stick Chart, Marshall Islands, c. Early Twentieth Century

A sailor's map made of sticks and shells

Maps can be woven and chiselled in rock. They can be spun in the air with your fingers, as you direct a stranger, 'go that way, then left'. There are many traditions of fashioning maps with materials picked up from the ground or off the beach but none so enchanting as the stick maps created by Pacific islanders.

This 'stick chart', which is just under 70 square centimetres (4 ⅝ square inches) in size, is the product of its place. It is fashioned from the ribs of palm fronds and cowrie shells and comes from the Marshall Islands, in the northern Pacific. The shells represent islands and for those able to unlock its secrets, the map contains a lot of information, including direction, distance and location.

The Pacific ocean is big. All the world's continents could fit inside its vast horizons. The islands that dot it are tiny, often low lying and a long way from each other. Finding your way in the Pacific is an epic challenge. All Pacific islanders have their own traditional ways of navigating it, often using the stars and knowledge about the motions of the sea. Each group of islands is different. Even by Pacific standards, the Marshall Islands are remote. They are 3,200 kilometres (about 2,000 miles) from any continent and all of them are small. Many are just the size of a few football pitches. This nation of sandy tropical islands is spread over nearly two million square kilometres (772,204 square miles) of ocean.

The stick maps of the Marshall Islands' sailors are beautiful but mysterious. A lot of their secrets will probably never be revealed. The maps were small enough to carry on the slender but robust canoes used in these waters, canoes with stabilizing floats called outriggers. Perhaps some voyagers did take them along but if we look again we can see why this was unlikely. This is more sculpture than pocket device: it's fragile and wouldn't last long on a cramped canoe.

What we have here is an object for study and learning. But it's not a public object, to be shared widely and equally understood by all. It is something unique, personal even, and some of its meanings may be known only by its creator. Hundreds of years of experience, passed from generation to generation, have gone into its creation. It is a summation of a particular patch of ocean, a reminder of the sea, to be handled and committed to memory before a dangerous voyage.

There are many types of stick map from the Marshall Islands. Some are flat rather than three-dimensional and contain complex representations of wave currents, turbulence and wind patterns. Others are like this one, sculptural and multi-dimensional. All of them were allied with hands-on knowledge. This is why we will never fully understand them: because we will never know how to feel the ocean like these sailors did. The skills of 'wave piloting' can only be acquired at sea. Young sailors would be blindfolded and left to their own devices so they would get to know the sensation of different swells, waves and ripples and what each could tell them about location and direction. Even as experienced mariners, islanders would shut their eyes and feel through their boat the sea's patterns and where to paddle for land. In this way the surface of the water was turned into a map, a microcosm that revealed the macrocosm.

Many ways have been found to map the sea, and many use objects that lie around and about its shore. Our main map uses palm fronds and shells but I want to show you another, this time carved from wood. A world away from the warm waters of the Pacific, the indigenous peoples of Greenland had to navigate icy seas. They too had small canoes but, unlike the Marshall Islanders they had a long coastline to help them find their way. It is this coastline that is being mapped on the two maps shown below. The two maps are two carved objects, each representing the coast north of the village of Ammassalik, in the south-east of Greenland. The one on the left shows headlands and fjords – it even has a groove showing where you can carry a boat overland between fjords – while the other shows a line of offshore islands. Just like the Marshall Islanders' maps these were objects of reminiscence and study rather than go-along aids.

There is a romance and beauty to these cartographic sculptures, but it comes as no surprise that neither the Marshall Islanders nor the Greenlanders use them today. They are relics from the past and while we know some things about them much has been lost. The decline of these mapping traditions can be put down to the sheer effort they required but also the predations of the modern world. The Marshall Islands endured German, Japanese and American occupation and then were turned into a nuclear testing site. The country's Runit and Bikini atolls were subjected to dozens of nuclear explosions: from 1948 to 1958 they were pounded with sixty-six nuclear devices. Some of these bombs were colossal: one dropped in 1954 had an impact 1,100-times larger than the bomb dropped on Hiroshima. During these years, these two miniscule islands accounted for more than fifty per cent of the world's nuclear fallout and many of the Marshall Islands became uninhabitable.

It's a shocking culmination of a story about an enchanting object. The entry of these remote places into the modern world was traumatic. The tourist brochures that lure tourists to the Pacific tell a truth but not the whole story: Pacific islands have palm trees and blue lagoons but they are not paradise. They are places of calamity and loss. There have been efforts in the Marshall Islands to revitalize the old ways, the old knowledge, but there is no way back. There is no more reason for sailors in the Pacific to use 'traditional methods' than there is for sailors anywhere else. Even less so: for finding your way across the world's biggest ocean safely leaves zero room for nostalgia.

Yet, for reasons that are hard to pin down, nostalgia persists, coming in with each tide. These maps evoke a poignant fascination, both for us and the modern islanders themselves. They speak of things distant and wonderful and of a relationship to the sea now gone which was as intimate as the touch of skin.

Two carved maps, Greenland, 1885

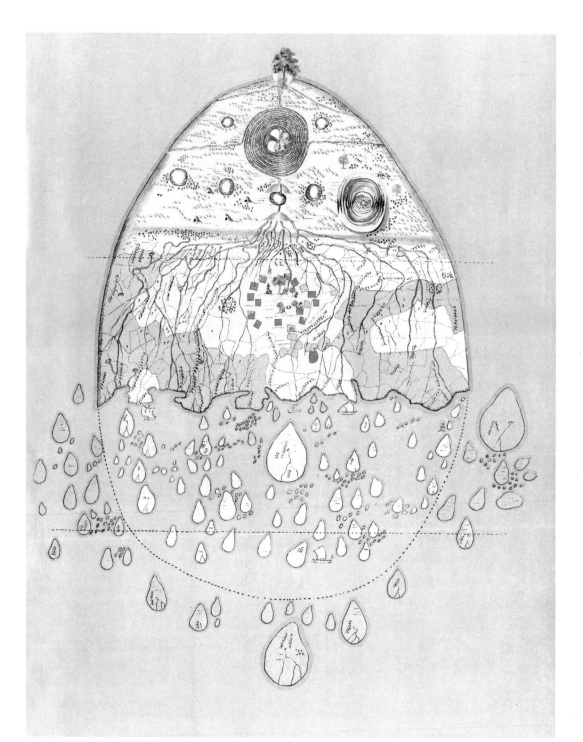

Map 9

**'A Burmese Map of the World',
reproduced in *The Thirty-Seven
Nats: A Phase of Spirit Worship
Prevailing in Burma* by Sir Richard
Carnac Temple, 1906**

The shattered tear drop: a journey to another shore

The world is a shattered tear drop, coursed with rivers and breaking apart into tear-shaped islands and, on the top of all this, stands a single green tree. This map takes us on a journey. It is not just a journey across miles, across physical distance, but to a different worldview. We have to leave preconceptions and certainties far behind and walk towards a different shore.

This is a map of south Asia. It's also a Buddhist map: a divine vision. The map made its first appearance in print in 1906 when it was reproduced in a book compiled by the English governor of Mandalay and folklore enthusiast, Sir Richard Temple. Despite his colonial pedigree, Temple's studies are a reputable source and continue to be cited by scholars. It is equally true that this map is a mysterious object: its creator is not known and nor are all the meanings that are buried within it.

It covers a territory that in ancient Indian sources is called *Jambudvīpa*, meaning both the island and the continent of Jambu. Jambu is a type of plum tree and it is this tree we see planted at the very top, the apex, of this oval realm. Below are representations of mountains, including the Himalayas, and seven great lakes, shown within a swirling circular patten. The lakes are folded together like petals. In the central lake grows a sacred lotus, the spring and source of the world's rivers. These rivers swirl in a labyrinth and then pour down across all the lands. There is another spiralling ring in the upper part of the continent. This is Mount Meru, a five-peaked mountain sacred in Buddhist and Hindu cosmology.

Heading south of the mountains we begin to get into less mystical territory. Temple tells us that in these lower portions we find the 'the world inhabited by all the human beings of whom the Burmese have any experience'.

The most important of these places is India, the fountainhead of Buddhism and Burmese spirituality. Right at the heart of the map is another tree. This is the Bodhi tree (the tree of awakening), under which Siddhartha Gautama gained enlightenment and became the Buddha, the 'awakened one'. If you peer really closely you can see the Buddha sat beneath it, and a few other figures to the side, which may be Buddhist monks. Around this compound are the scared sites of the Buddha's birth and life story. These are all in India and represented by small red squares. Reportedly there is also a snake playing a mandolin. Can you spot it? I keep looking.

This far down the map we arrive at known pilgrimage sites and we also enter the modern world of nations. They are shown in different colours with clear borders. The middle one, in which the Bodhi tree stands, is surely India but none is named. There is little in this map that is straightforward. Temple suggests that, among other things, the map is an 'attempt to copy a coloured European map of the seventeenth century'. The borders, along with the way the mountains and rivers are drawn, do hint at the adaption of a European mapping tradition. Another European touch are the two faint dotted lines of latitude, which run right the way across the map. But even these recede from easy interpretation. It is not clear what latitudes they express, if, indeed, that is what they are intended to be.

The teardrop comes apart as we enter the ocean. The sea is decorated with islands of all sizes. This is neither whimsy nor a cartographical reverie. We see boats, plying their trade. Temple reads all this through the lens of what the Burmese did and didn't know at the time, and writes that 'all this is "natural" geography. All the seas known to the Burmans are to the South, and the land of great mountains is chiefly to the North.'

So why does the tear drop shatter? Why is its complete outline drawn in as a dotted line, and why within that line are the islands yellow and pale green but outside it they are all pink? This suggests a lost wholeness, but also inner and outer islands. The islands all have the same teardrop shape but they are not identical; some have features, such as mountains and rivers, though none has any sign of habitation. They seem like places rumoured of but little known and far away.

Temple reproduced another map alongside this continent-sized teardrop, and the contrast is striking. Serene colours and peaceful symbolism are replaced by ghastly red and jet black and scenes of torture and suffering. This is a map of hell, and appears to be seventeenth-century in origin. The Buddhist hell, called Naraka, is part of the cycle of rebirth. Accumulated bad actions in previous lives mean you might get born in hell and stay there until your bad karma is paid for and expended. Buddhist hell is long-lasting but not eternal: eventually you get released into a higher realm. The Burmese version reproduced on this page shows us the various caves that make up one level of hell, as well as the main entrance.

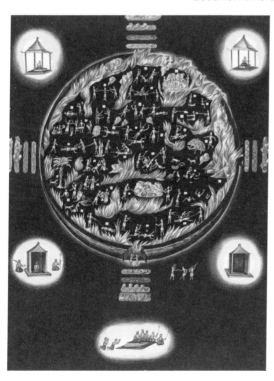

It is an arresting image for the odd reason that it is so familiar. Similar representations of hell can be found across the world, including in many European paintings. In comparing the two maps, it is hard not to conclude that the human imagination has a rather limited repertoire of the hereafter – heaven and hell always look pretty similar; clouds and fire – but a fantastically diverse grasp of the real world.

'Hell according to the Burmese', reproduced in The Thirty-Seven Nats: A Phase of Spirit Worship Prevailing in Burma by Sir Richard Carnac Temple, 1906

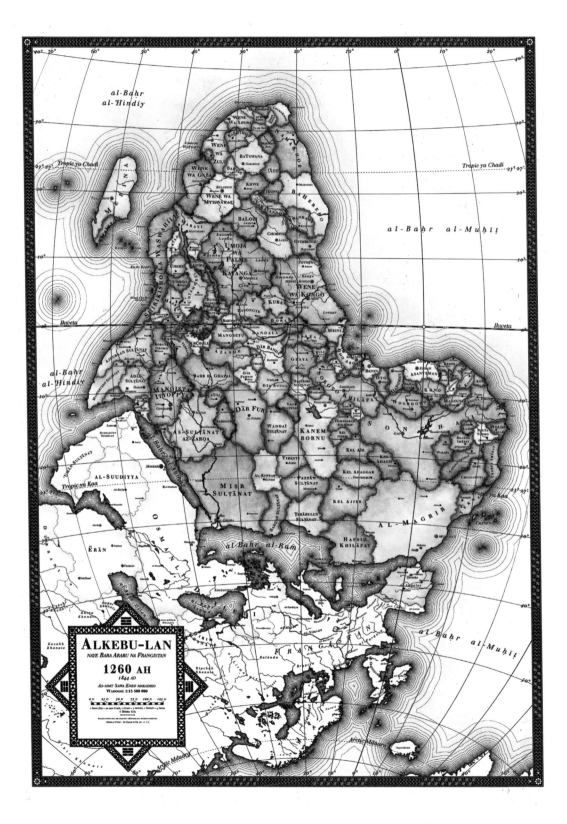

al-Bahr al-'Hindiy

Tropic ya Chadi

al-Bahr al-Muḥiṭ

Ikweta

Ikweta

al-Bahr al-'Hindiy

al-Bahr al-Rūm

AL-MAGRIB

Tropic ya Kaa

Tropic ya Kaa

al-Bahr al-Muḥiṭ

Arctic Mduara

Arctic Mduara

ALKEBU-LAN

NAYE BARA ARABU NA FRANGISTAN

1260 AH

1844 AD

AS-SIMT SAWA ENEO MAKADIRO
WADOGO: 1:15 500 000

Map 10
Alkebu–lan, created by Nikolaj Jesper Cyon, 2011

If Africa had not been colonized

The borders and states of Africa arose, in large part, from the ambitions and imaginations of European colonizers. The question 'what would Africa look like without colonialism?' is impossible to answer yet necessary to ask.

By 1914, Europeans controlled ninety per cent of Africa and its borders today reflect this dominance. What if that had not happened? 'What if?' is a well-known question in history. It leads to 'counterfactual history'; the history of what didn't happen but might have. Things don't have to be the way they turned out.

This map was created by the Swedish graphic artist Nikolaj Jesper Cyon and he based it on historical empires and ethnic and linguistic regions. It's a snapshot from the year 1844. Cyon explains that he 'first and foremost looked at historical states', by which he means 'territories with a centralized government during the point they controlled the largest areas during the time frame of 1300–1844 AD'. In places where a centralized state did not exist Cyon 'looked at names of cultures that have inhabited an area' and, he says, as a 'last resort', he used 'the names of the language groups that existed in the area'.

Cyon is guessing that without colonial control, Africa's states would have arisen in the same way they did in Europe. Borderlines in Europe are not straight: they wriggle around, going in and out of valleys and making odd twists. They reflect the complex pattern of where, say Italians, or Germans, live. By contrast, the current map of Africa has a lot of straight lines. The 1,276-kilometre (793-mile) border between Egypt and Sudan is an arrow shot across the desert, imagined and then drawn by British colonial officials. Take away those officials and what might you have got? Something much closer to what we see here, an irregular border that reflects the vagaries of centuries of history.

The first thing we notice about this map is that it's upside down. This flip was inspired by Islamic mapping traditions which sometimes did the same and pushes us to rethink and re-see Africa. By turning things on their head the map-maker is drawing hidden assumptions into the light. The keen eyed may also notice that the *ikweta* – Swahili for equator – bisects a new central meridian. A meridian is the line that time is measured from. The Greenwich Meridian is the world's meridian: the line that runs through London is the point from which every other time zone is gauged. Head east of London and you are heading in advance of Greenwich mean time, and you need to turn the hour hand of your watch forward; go west of London you need to wind it back. In Cyon's map the meridian goes through Timbuktu in Mali.

It's a good choice. Timbuktu was once a great trading hub and centre of learning. In the medieval era its university had 25,000 students. In a nod to this heritage, this map makes Timbuktu the temporal centre of the world; the city to which all the planet's time zones refer.

All the countries on this map have their own stories. I'll highlight just a few here. At the bottom of the map Spain is labelled *Al–Andalus* and

Extract of updated map, courtesy of the artist

is part of the nation of *Al–Maghrib* (Arabic for 'the West'). It's a larger version of Morocco and adapts its name; one of the names Moroccans give their country is 'the Kingdom of the West'. As elsewhere the map extrapolates a real history. The Islamic conquest of North Africa did extend over nearly all the Iberian Peninsula (Spain and Portugal). In a world without the European rise to power, it is likely this would have still been the case.

The borders on what we might call Arab Africa, or 'North Africa', are the most recognizable lines on the map. If we look at the *Misr Sultanate*, at the bottom left of the map, it's clearly Egypt. *Misr* is what Egyptians call Egypt and its frontiers here are familiar. It's a reminder that Europeans were not the only foreigners to have an impact on Africa. Many centuries before the Europeans, Arab armies also changed borders. By 698, Arab led armies had taken most of North Africa, reshaping countries and traditions.

Another story can be told about the island up at the top left. This is the Kingdom of Merina, which today we call Madagascar. Merina existed as an independent kingdom from 1540 to 1897. The last Merina sovereign, Queen Ranavalona III, came to the throne at the age twenty–two. Following a French invasion, she was exiled to Réunion and later Algeria, where she died at her villa in 1917 at the age of fifty–five. That was the end of Madagascar's royal line and the Kingdom of Merina sank from memory.

Each of these imagined countries has a story but that does not mean that they cannot be questioned. A lot of them assemble disparate peoples. You may be able to see a country in the middle and left-hand side of the continent (with a purple tinge) called Buganda. The name sounds familiar because the kingdom of Buganda lent its name to the country of Uganda and Buganda still exists in the form of Uganda's largest region. However, on this map its extent is huge: it reaches right across Uganda and neighbouring countries. This *Buganda* absorbs many tribes, languages and potential nationalities. The decision to name the whole place after just one part of the whole is controversial. Counterfactual mappings and counterfactual histories show possibilities rather than probable outcomes.

Not far from Buganda, down a bit and further left, with a grey-black tinge, we may notice another familiar country with an unfamiliar name. *Mangista Ityoppya* means Empire of Ethiopia. In 1844 Ethiopia was an independent country and, with some interruptions, it remained so. This sovereignty had to be fought for. A famous Ethiopian painting seen here shows one of the key moments in that national defence. At the top and sitting on a white horse is Saint George, the national saint of Ethiopia, as he had been for a thousand years. He is exhorting the Ethiopian army, shown on the left, against a force of invading Italians. This is the Battle of Adwa, which took place in 1896 and was a famous victory for the defenders. It's a win marked today as a national holiday in Ethiopia and it's a reminder that the map of Africa is not just a European or Arab creation but is full of African stories too.

St. George, patron saint of Ethiopia, overseeing the battle of Adwa

1926

1926

1760-1762

1809

1743

Arkhangelsk

Saint Petersburg
Novgorod   Vologda
Suzdal
Moscow
Smolensk   Riazan
Kylv

1476

1552

1522

Omsk

Krasnoyarsk

Yakutsk

Okhotsk

Irkutsk

1914, 1932

1858

Manchuria
1900-1905

Khabarovsk

1875-1905

Warsaw
Ternopil
1809-1815
Wallachia and
Moldova
1835-1856

Astrakhan

1859

1860

Kotor
1806-1807

Ionian Islands
1800-1807

Tbilisi

Buxoro

Toshkent

Yining (Qulja)
1871-1881

Vladivostok

Lüshunkou (Port Arthur)
1858-1905

South Caspian
1723-1732, 1907-1921

Sagallo
1889

1867

1761-1762

1761-1867

Kodiak

Sitka

mchatsky

Fort Ross
1811-41

Fort Elizabeth
1818

Map 11
Territorial Expansion of Russia, 1300–1945

An empire made visible

Colonialism does not come in one flavour. A lot of people think of it as the conquest of distant lands and empires as fragmented assortments of foreign territories. Russian colonialism and the Russian Empire are different. Russia simply expanded and absorbed its neighbours.

This has proved an enduring model. We look at the map of the world today and see Russia reaching all the way to Vladivostok. Few pause to wonder how Russia got so large. Our map shows that Russia once went even further. Pushing east, Russia became a major colonizer of North America. Russian expansion along the Pacific coast was driven by the pursuit of the pelts of bear, beaver and other wild animals. Russian hunters swarmed across the Bering Strait, and in 1784 Russia's first permanent settlement in Alaska was established at Three Saints Bay. Soon missionaries began to arrive, converting native tribes. The Russian–American Company, which took over the fur trade, established a capital at New Archangel, and then pressed on, extending Russian settlement down the coast, as far as northern California. This is where Russia's Fort Ross, (pictured opposite) built in 1812, still stands. Today this vestige of Russian America is a remote museum. .

Russian frontiersmen were hoping to turn a profit and when overhunting led to a collapse in animal skins they were on borrowed time. The Russian court was impatient and eager to rid itself of burdensome possessions. In 1867 Emperor Alexander II offered Alaska to the United States for $7.2 million. The question of whether it was Russia's to sell or the USA's to buy did not occur to either party.

By far the largest and most important territories brought under Russian control were in Asia. These were places not just for hunters and trappers but for long-term settlement. Most are still in Russia. From 1582 to 1650, Russia invaded and subdued Siberia (part of the green area on our map). Russian armies overwhelmed the Siberian Khanate and reached the Pacific in 1639. Conquest was followed by the founding of new towns and cultural absorption, with Russian becoming the dominant language. Another expansion occurred from 1785 to 1830 and looked south to the lands between the Black Sea and the Caspian Sea and saw the taking of Armenia and Georgia, both coloured in yellow on our map. By 1829, Russia controlled all of the Caucasus. Later that century Russia annexed parts of northern China, then turned west, incorporating swathes of central Asia. The growth of Russia was unrelenting and changed the destiny of Asia.

Ukraine, like nearly all of imperial Russia's colonial possessions, did not achieve freedom with the communist revolution of 1917. Finland was the exception that proves this rule (it gained independence in 1917). Lenin and Stalin presided over another phase in Russia's colonial history. In 1922, Bolshevik Russia joined Belarus, Transcaucasia and Ukraine as the four founders of the Union of Soviet Socialist Republics, which lasted until December 1991. Although technically a merger of equals, there was never any doubt that this vast multicultural empire was ruled from Moscow.

The Russian model of colonialism by absorption confuses the divide between colonized and colonialists that was so stark in the British, Spanish and French Empires. Russia's expansion was, in many ways, less destructive. The transatlantic slave trade and the extermination of indigenous people seen in the Americas have no parallels in this history. The Russian Empire can, and maybe should, be judged less harshly than the empires of Western nations. Yet the 2022 invasion of Ukraine points to another viewpoint. For one of the consequences of absorption, or a rolling forward of borders over next-door lands, is a lack of recognition that any conquest has taken place and, hence, an almost impermeable sense of entitlement. The result is a firm conviction that contiguous lands are not real countries, but mere annexes of Russia. For pro-Putin Russians, it is almost impossible to imagine the attempted take-over of Ukraine as an invasion; it's more akin to taking back mislaid property.

For Russian ultra-nationalists, Ukraine is framed by its old imperial label of 'Little Russia', and the Ukrainian language is 'Little Russian'. We might also be reminded of the curious imperial decree from 1863 which banned Ukrainian-language publications on the grounds that 'no separate Little Russian language has ever existed, exists or can exist'. Banning something that you claim does not exist might seem unnecessary. It speaks of a combination of defensiveness and dismissiveness.

This map shows the waxing and waning of Russia. Most countries wax and wane but the scale and complexity of Russia's advances and withdrawals is unique. All nations have changing maps but the drama evident in Russia's map is non-stop. Today, along with the one in Ukraine, Russia is involved in numerous disputes with its neighbours. It is quarrelling with Japan over the Kuril Islands and with Moldova over Transnistria. In 2008 it invaded Georgia to seize the break-away regions of South Ossetia and Abkhazia.

Unlike the expansionist incursions of the past, many of these disputes are about making a claim on areas where Russian speakers live. In the language of geopolitics, Russia has become 'irredentist', an ideology that argues that ethnic compatriots must be included in a single, ethnically defined, state. To this end, Putin points to the oppression of Russian speakers as a defining feature of the countries he invades.

Yet, if the last century taught us anything, it was that countries can contain diversity. Russian speakers don't have to be Russian, any more than English speakers have to be English. In the twenty-first century attempts to 'rescue' ethnic minorities for the 'motherland' are a recipe for conflict.

Fort Ross, Sonoma County, California

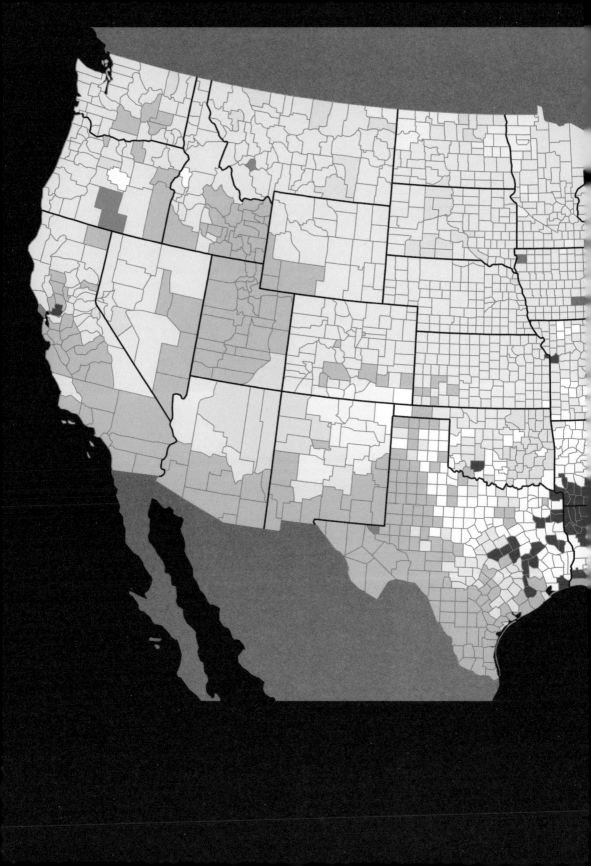

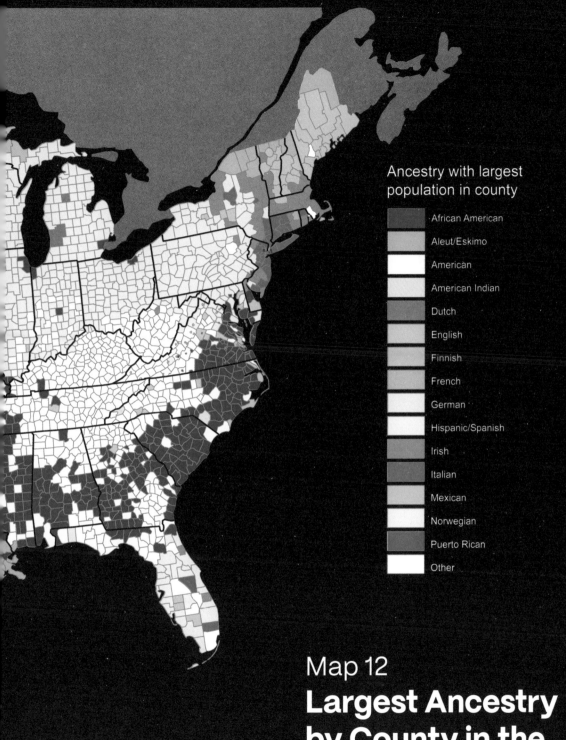

Ancestry with largest
population in county

- African American
- Aleut/Eskimo
- American
- American Indian
- Dutch
- English
- Finnish
- French
- German
- Hispanic/Spanish
- Irish
- Italian
- Mexican
- Norwegian
- Puerto Rican
- Other

Map 12

Largest Ancestry by County in the USA, 2000

The German USA

The expanse of light blue that dominates this map spreads out from the heartland of the USA and surges north, west and east. These are all places where German Americans are the single biggest ethnicity. There is a broad swathe in the south where this title alternates between Mexicans (light pink) and African Americans (dark purple), and up in the far northeast, English (light purple) and Irish (mid-purple) predominate.

The American census asks people how they would describe their ancestry and it is from this question that we get this map. People don't have to pick a foreign country. Some people – such as the majority in Tennessee and Kentucky – said 'American'. But the most popular answers refer to migrants.

The fact that Germany is such a major source may come as surprise. It is not one of the origin stories about which a lot of Hollywood films are made. About one million Americans speak German but they are spread thin. It is the second-most-spoken language in North Dakota, but few people live in North Dakota and only 1.4 per cent of them speak German.

The last century did not always make it easy to be a German American. Two world wars pitted the USA against Germany and, for a while, turned a founding identity into a hidden identity.

By looking at another survey of self-identified ancestry, the American Community Survey from 2022, we can add more detail. Those who said their ancestry was German are still the largest faction (41 million). Among other European heritage groups they are followed by the English and Irish (both at about 31 million). We also come across other sizeable European heritage communities who, like the Germans, we don't hear much about, such as Swedish (3.3 million), Dutch (3 million) and Norwegian (4 million). Other groups from the British Isles also feature strongly, namely the Scots-Irish (2.5 million), Scottish, (5.3 million) and Welsh (1.5 million). Almost two million said British. That surprised me. Escaping Britain was America's defining act.

How and why we pick the ancestors we identify with is complex. It is a choice. Very few people, in the USA or elsewhere, are a hundred per cent anything. We identify the bits of our heritage that we're proud of and that choice shapes who we say we are. Sometimes, the reasons for doing so may be no more profound than the hope that a particular ancestor might make us appear interesting. Perhaps, I should draw a veil over one of my relative's announcement that Pocahontas was one of my distant relations. But I can't help it. It's nonsense but it gave me a momentary thrill. Ancestry is interwoven with wishful thinking.

There are plenty of reasons why so many wanted to flee Germany. Before 1871 Germany was politically fractured and jobs, land and political rights were in short supply. This fragmented land was also wracked with political unrest. A wave of riots in 1848 expressed a widespread desperation and millions decided they'd fare better across the Atlantic.

German immigrants, Ellis Island, New York, 1920

From 1860 to 1890 Germans were the largest group of arrivals to the USA. They would have been encouraged by the knowledge that they were not the first; that this was a country with a long tradition of German settlement. In 1608 Germans joined the first permanent English settlement in the Americas, Jamestown, which was just one year old, and in 1620 they were active in the founding of the Dutch colony of New Amsterdam (today called New York). These 'Germans' brought with them specific skills, such as the manufacture of glass and tar. I use quote marks because, in these early years, people did not use the same labels we have today. Religious faith and language often meant more than nationality.

Pennsylvania has the largest population of German Americans in the USA and one of the oldest German settlements, Germantown founded 1683. It is now surrounded by the city Philadelphia and is one of seventeen Germantowns in the USA.

The eighteenth century saw substantial and diverse German migration. One source were the mercenaries the British brought over to fight on their side. About 30,000 'Hessians' were hired. Even though they were on the losing side, some of them stayed. Another group, hidden in the figures, are German Jews. Many tens of thousands emigrated in the nineteenth century. With the rise of the Nazis a new wave began and about 100,000 came to America, especially to New York. Most read and spoke German and had a big influence on American culture. But, as we have seen, ancestry is, in part, chosen. Given the choice, many German Jews identified themselves as Jewish Americans rather than as German.

By sheer weight of numbers, the size of the American population that can trace its roots back to Germany makes it the sleeping giant of American identities. We can pick out many American customs that have a German stamp. Kindergartens are one; another is the Christmas tree, and food like hot dogs and hamburgers. Many of the best-known American brands were founded by Germans: from Boeing (William Boeing's parents came to the US in 1868) to Levis (in 1853 Levi Strauss moved from Bavaria to San Francisco).

There is a footnote to this story. Germany *was* a country of emigration. People left it. Now the opposite is true. People want to come. Today about twenty per cent of the population of Germany is foreign born. This transition marks the shift from being a country that lacked opportunities and rights to a country that has both. Once upon a time, not so long ago, large numbers of Germans were able to take advantage of the very things that now lure others to Germany.

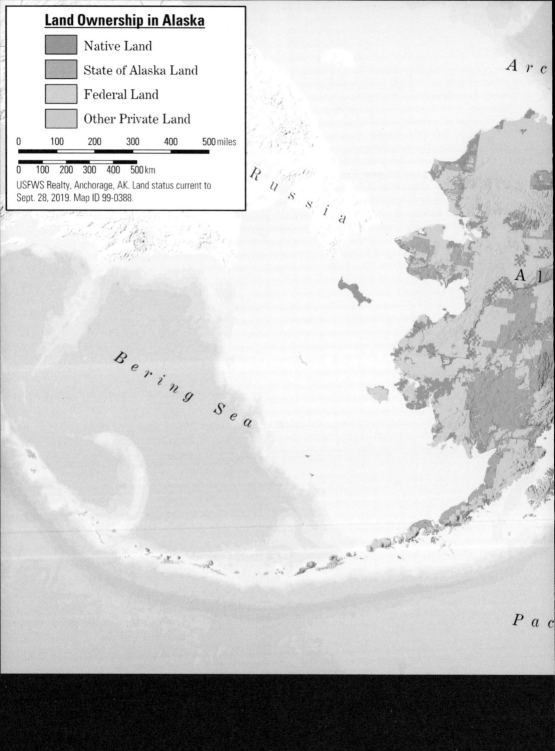

Land Ownership in Alaska

- Native Land
- State of Alaska Land
- Federal Land
- Other Private Land

| 0 | 100 | 200 | 300 | 400 | 500 miles |

| 0 | 100 | 200 | 300 | 400 | 500 km |

USFWS Realty, Anchorage, AK. Land status current to
Sept. 28, 2019. Map ID 99-0388.

Russia

Bering Sea

Arc

Al

Pac

O c e a n

C a n a d a

a

O c e a n

Map 13
**Land Ownership
in Alaska, 2019**

Who owns Alaska?

When the US bought Alaska from Russia in 1867 it was twice the size of the thirteen original American colonies. Almost all Alaskans were indigenous Alaskans. Their claim on this land was not even considered. That claim is starting to be recognized today but the stand-out feature of this map is just how much of Alaska is in government hands – all the bits in green. All those in blue are owned the State of Alaska.

One of the stereotypes about the USA is that it is a country of small government, a nation of free-markets and private wealth. The latter bits are true but 'small government'? Not really. The government is the major land-holder, especially in the west and in Alaska. Alaska isn't even the state with the largest amount of federal land. That title goes to Nevada, eighty per cent of which is owned by the federal (i.e. central) government; the figure is sixty-three per cent in Utah, and sixty-two per cent in Idaho. It's sixty-one per cent in Alaska, the largest of all the states (Alaska is six times bigger than the United Kingdom). Overall, the federal government owns twenty-seven per cent of the United States. Many state governments also have huge holdings.

With great power comes great responsibility. In particular, the long overdue restoration of land to the indigenous peoples. For nearly twenty years after the US paid Russia $7.2 million for Alaska in 1867, it remained closed to private buyers or any other kind of redistribution. The territory was then opened to mining companies and homesteading, but its remoteness and hard climate kept claims down and only a fraction of the land was ever bought. When Alaska was upgraded from a Territory to a State in 1959 only half of one per cent did not belong to the federal government.

Becoming a state altered the balance of power. Alaskan politicians wanted, and got, much more control. The State was handed a fat slice: twenty-eight per cent of the land. It took some time before native claimants got their turn. The 1971 Alaska Native Claims Settlement Act handed over – perhaps 'handed back' is more accurate – 17,806 square kilometres (6,875 square miles). It may sound a lot but Alaska covers about 1.7 million square kilometres (0.66 million square miles).

Today the federal government remains the biggest player and still manages land set aside both for the protection and the exploitation of nature. This includes national parks and forests, like Kenai Fjords, where moose, wolves and bear roam free, as well as nine military installations, and the North Slope National Petroleum Reserve, which covers some 93,000 square kilometres (about 35,900 square miles).

Nanieezh Peter and Quannah Chasing Horse address the Alaska Federation of Natives Convention, 2019

This is not a frozen map and we're grateful to Scott McGee at the US Fish and Wildlife Service and Jeffrey J. Brooks for providing us with an up-to-date version at time of writing. There has been a concerted push to return more land to indigenous peoples.. The Alaska Land Transfer Program oversees the largest land transfer effort ever undertaken in the United States. It's a complex process involving surveys of tens of thousands of miles of tracks, legal verifications and agreements. So far the 'entitlement' of the combined interests of 'Alaska Native Corporations, Native Allottees, and the State of Alaska' has been worked out at 18.5 million hectares (45.7 million acres) (the pink land on our map), roughly the size of the state of Washington.

No discussion of land in Alaska today makes sense without talking about oil. The 1971 Settlement Act was not just about land but also about oil. The Act paid over money to native people: precisely $962.5 million. But half of this was to be paid from royalties on oil production. Oil was the key factor in getting speedy resolutions of land claims. Byron Mallott, who was leader of the leader of the Kwaash Ké Kwaan clan as well as Lieutenant Governor of Alaska, explained that the settlement of native claims 'was all about the economic value of oil'. Native claims have been accepted on the basis that native people won't stand in the way of oil production. This was once seen as a win-win. But climate change has changed that calculation. Today it looks like there are no winners.

Today many Alaskans are questioning their oil-based economy. Alaska is warming at two to three times the rate of the global average, which is having a serious impact on its landscapes and inhabitants' livelihoods. Numerous villages have had to move, as the ground below them melts and disappears. Even the language is changing. Indigenous words like *tagneghneq*, used to describe dark, dense ice, have become obsolete as the permafrost gives way and a once solid landscape breaks up.

Native Alaskans were among the first to warn of climate change. Patricia Cochran, director of the Alaska Native Science Commission, says that 'it has taken science a very long time to catch up to what our communities have been saying for decades'. She continues that for 'at least the last forty or fifty years, our communities have noticed the subtlest of changes' and that they were:

seeing the signs of climate change long before researchers and scientists started using those words. Climate change is more than just a discussion for us. It is a reality. It is something that we live with and face every single day – and have for decades.

The development of environmental activism among indigenous youth is particularly striking. Our photograph shows Nanieezh Peter (aged fifteen), on the left and Quannah Chasing Horse Potts (aged seventeen) on the right, speaking in support of declaring a state of climate emergency during the 2019 Alaska Federation of Natives Convention. They look determined but anxious, uncertain of the reception their message will get. A lot of people, both natives and settlers, are not onside: the oil business has offered serious money over the decades and individuals and whole communities have been beneficiaries. Not everyone wants to rock the boat.

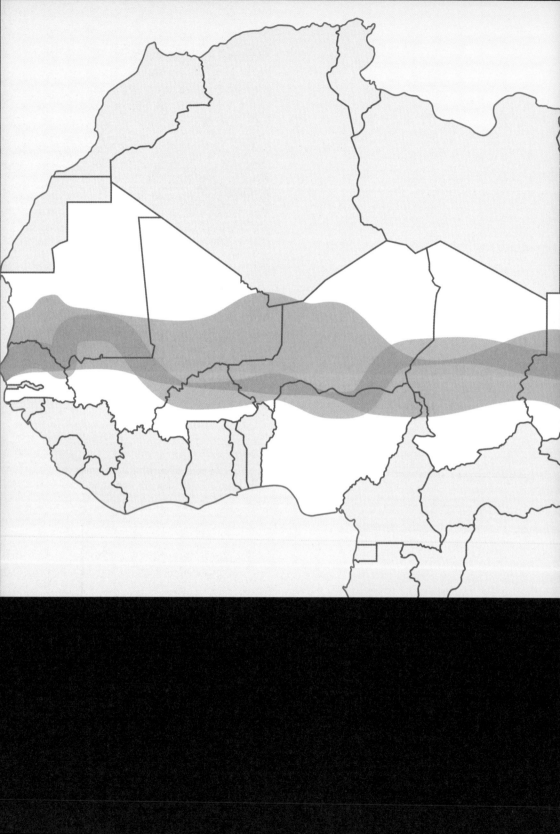

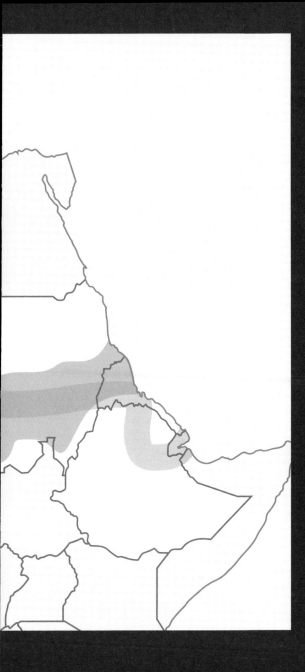

Map 14
**Africa's Great
Green Wall, 2021**

A world–leading environmental plan
from Africa

Severe land degradation affects 168 of the world's countries. Soil is a complex ecosystem and as temperatures rise it begins to die and turn to dust. The consequences are obvious: dead soil means crops don't grow, so people go hungry. This is happening while the planet's population keeps rising. In 1950 it was two-and-a-half billion, it is now eight billion and is expected to hit ten billion in 2050.

All sorts of solutions are being rolled out in a bid to save our soils. Across China reed and grass cages are woven to keep topsoil in place; in Spain a new national strategy is promoting less-intensive farming. By far the biggest scheme is to be found in Africa. From Senegal to Djibouti – that's about 6,500 kilometres (4,040 miles) and not far short of seven times the length of Britain – a broad swathe of trees is being planted, keeping the desert at bay and protecting farmland. This is a simple map but, in terms of the number of lives it could benefit, it might be the most important one in this book. The Green Wall is the thick green line on the map, the brown band represents the geographical region it runs through, the Sahel, a zone that lies between the deserts of the Sahara and more humid, wetter lands to the south. The first tranche of participating countries is also highlighted.

It is a work in progress. By 2023 only eighteen per cent had been planted, although they still cover 18 million hectares (around 44.5 million acres). The achievements are real and so are the challenges. The Green Wall travels through uncertain times and many of the world's poorest and most unstable countries. 'The desert is a spreading cancer,' Abdoulaye Wade, Senegal's president and one of the wall's standard bearers, explained, adding 'we must fight it. That is why we have decided to join in this titanic battle'. Back in 2007, the heads of the countries highlighted on the map (going from left to right) – Senegal, Mauritania, Mali, Burkina Faso, Niger, Nigeria, Chad, Sudan, Ethiopia, Eritrea, Djibouti and Somalia – got together to set it in motion.

Since the late 2010s, there's been both a rethink and an expansion of the wall's ambitions. Many of the trees planted in the early years died. It turned out that a system relying on outsiders to plant the trees was too top-down. It's better to work with local farmers, many of whom have cheaper and more effective solutions to problems, such as using simple water harvesting techniques and protecting trees that emerge naturally on their farms. This new approach is central to the new phase of Africa's Great Green Wall and now twenty-one countries have joined the reborn and re-energized project.

The Green Wall is not much known outside Africa although some of its partner institutions are. One is the International Olympic Committee which is contributing an Olympic Forest, and some 600,000 additional trees to the Wall, in Mali and Senegal. The eco-friendly internet search engine Ecosia is also a partner, making extensive contributions of trees. Partnership but also innovation characterizes the project. Existing and

At the edge of the desert:
the Sahel in Mali

traditional knowledge is being put to use but so too are inventions, such as the small planting boxes called Groasis (as in, 'Grow-oasis'). This Dutch design cleverly traps and retains water. It is claimed to be ninety per cent cheaper and use ninety per cent less water than conventional irrigation methods like drip feeding water from pipes. In 2023, Nigeria's National Agency for the Great Green Wall signed an agreement for Groasis box production to be located in Nigeria. A Nigerian Government website explains that it will 'accelerate Nigeria's Tree Planting campaign in the [Great Green Wall] belt, boost economic activities, create thousands of new jobs'; in aid of 'transforming the Great Green Wall belt into an economic hub'.

As this messaging makes clear, the Great Green Wall is understood to be an economic lifeline, one which will hopefully slow down the exodus of people from the countryside that is swelling African cities. With dying soil and dying farms comes depopulation which, in turn, leads to an unmanageable growth of cities and to people becoming dependent on food imports. The Wall will help cut this dangerous chain. There are also political benefits. Hunger and poverty create ideal conditions for the growth of extremism. Over recent years, Islamist terrorist groups have been thriving in the increasingly barren and desperate lands through which the Wall is going. Improvements in land quality are better than armies and bullets in combatting extremism.

The Great Green Wall is unique in scale and ambition, but similar ideas have been around for a while. Although smaller, the Great Plains Shelterbelt project in the USA tried to do something similar. From 1934 millions of trees were planted to help stop soil erosion. The idea was that trees would provide a windbreak and stymie the creation of Dust Bowls. By 1942, 220 million trees had been planted and covered 48,000 square kilometres (18,533 square miles). The project was judged a success, although its legacy has been rather frittered away. When they died, many of the trees were not replaced and the 'shelterbelts', though they hang on in patches, are not what they were.

There is a lesson in the story of the Great Plains Shelterbelt: creating environmental protection is one half of a solution; the other is maintaining it. This appears to be understood by the Chinese. China's Great Green Wall (named the Three-North Shelter Forest Program) is currently hard at work holding back the Gobi Desert. China's Great Green Wall began in 1978 and is now almost complete, stretching some 4,500 kilometres (2,800 miles). India has a smaller equivalent in the Aravalli Green Wall Project. It has been designed as a 1,400-kilometre (870-mile) forest 'belt buffer' around the Aravalli Mountain Range and protects farmland in the states of Haryana, Rajasthan and Gujarat.

Green walls are on the rise. Africa's is the most impressive; an example of environmental collaboration and community involvement from the poorest nations on the poorest continent. Richer nations should take note and learn.

比例尺 1：60 000 000

Map 15
Vertical Map, Hao Xiaoguang, 2013

New maps for a new superpower

World power is shifting, and cartography is a servant of power. This Chinese map puts Asia centre stage. It dominates the map. The Americas are shunted to the edges and radically severed, with North America on its 'side', creeping around the top of the map (see opposite), and South America slipping up from the south.

In China, a more conventional redrawing is to simply shift the world eastwards, making east Asia central. Either way China begins to look like the 'middle country' its Chinese name proclaims it to be. The Chinese for China is *Zhōng guó*, which combines two characters, 中 (middle) and 国 (country or kingdom), hence 中国, Middle Country. For thousands of years Chinese scholars had explained that China sat at the centre of the world and was the most civilized of all the nations. The last two centuries knocked the confidence out of this Sinocentrism. Today, it is returning and maps are a vital expression of the new self-belief. They make a tangible claim on the world. Patriotic speeches and music inspire and can fill the air with applause, but a map can move mountains.

The Chinese Government has got into the habit of publishing cartographical provocations. The one we have here is intriguing, but I don't find it controversial since I think the re-orientation it offers is both imaginative and a useful corrective. The map's creator, Hao Xiaoguang of the Chinese Academy of Science, makes the point that it shows far more accurately than horizontal maps, how close countries in the northern hemisphere are. The northern half of the planet is where the great majority of the world's population live so it's useful to see, as we can do on this map, that a flight from Beijing to New York would be far quicker going over the Arctic Ocean (11,000 kilometres/6,835 miles) than across the Pacific (19,000 kilometres/11,806 miles).

Hao Xiaoguang has created a map that is challenging but impressive. If we are looking for Chinese maps that are simply provocative then a good example was released in 2014 by China's Ministry of Education, which claimed most of the Pacific for China. Numerous sovereign countries were roped into a vast realm that gobbled up Hawaii and coasted down the western coast of the USA and Mexico. The President of Micronesia at the time, Manny Mori, called it 'cartographic rape'.

Chinese maps are pushing boundaries. After a Chinese Ministry official announced that 'the study of what constitutes Chinese territory is ongoing', talk of historical evidence that at least some of Antarctica was Chinese, began to circulate. Slices of Antarctica are currently claimed by Argentina, Australia, Chile, France, New Zealand, Norway and Britain. It's a long list of countries, and many of them are nowhere near Antarctica. It's not surprising that China, the world's new superpower, might want to throw its hat into the ring.

The hope that China can take a slice of the pie may explain why Antarctica looms so large in this map. The position of both poles is its standout feature. They are dragged out of their conventional lowly role, at the very top and bottom of the world. The continent of Antarctica, usually a toothy residual rim on world maps, appears in all its white empty

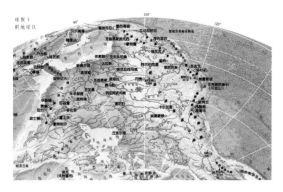

North America, on its 'side', creeping around
the top of the map, Hao Xiaoguang, 2013

glory and the Arctic, the North Pole – the blue sea circled with land that sits just above Asia near the top of the map – is shown as what it is likely to soon become, open water.

The most alarming thing about this vertical map is the way it represents the Poles as accessible and exploitable. It is becoming hard to believe that explorers searched so long and for so many decades to find a way through the endless horizons of year-round ice that once locked in Canada's north. For centuries mariners risked their lives hunting for the fabled Northwest Passage. It took Roald Amundsen three years, from 1903 to 1906, to become the first person to navigate it successfully and even then it was touch and go; at times the water Amundsen sailed on was only three feet (less than a metre) deep. Now look at it. On this map the whole of the Arctic is shown as blue, navigable water. Today the Northwest Passage can be sailed all year round. Cruise ships take thousands of tourists on leisurely reconnoitres of Amundsen's route.

Ice does still cover much of the Arctic. Hao Xiaoguang made the perfectly legitimate decision to only show ice when it's on land, such as over Greenland or Antarctica. Yet by placing the blue Arctic so centrally he showcases its emerging status as a trade route. The first transect of the Arctic Ocean was undertaken in 2004 as a collaborative effort between two icebreakers, the Canadian *Louis S. St-Laurent* and the US *Polar Sea*. No ordinary ship could have made the journey. But every year the ice gets smaller and thinner. To date the Summer ice cover lost, compared to the average Summer cover from 1981 to 2010, is about the size of Alaska.

There is still a lot of uncertainty about when a Transpolar Sea Route will become viable. There need to be safe, ice-free conditions for ordinary cargo ships. In other words, it needs the end of the North Pole as we know it. Some say it will be as early as 2030; other estimates push this back to 2050. Shipping companies are already incorporating the route into their long-term planning. The Transpolar Sea Route will revolutionize world transport. Travel times for goods taken between the east of Asia and the Atlantic will plummet and shipping via the Suez and Panama canals will shrink.

The Arctic has more than one type of magnetic attraction. It has been estimated that about twenty-two per cent of the world's undiscovered and recoverable resources – oil, gas and valuable minerals – are in the Arctic Circle. It may soon be witness to a twenty-first-century gold rush. The same fate may also await its southern twin, Antarctica.

Some will look at this map and see an Asian vision of the future. It is. But look again and something else is going on. The poles, so long overlooked, are being shoved in front of us as zones of economic potential. The remotest places on the planet are within our grasp. It is an uncomfortable moment. The world stands fully revealed and there is nowhere that will be left in peace.

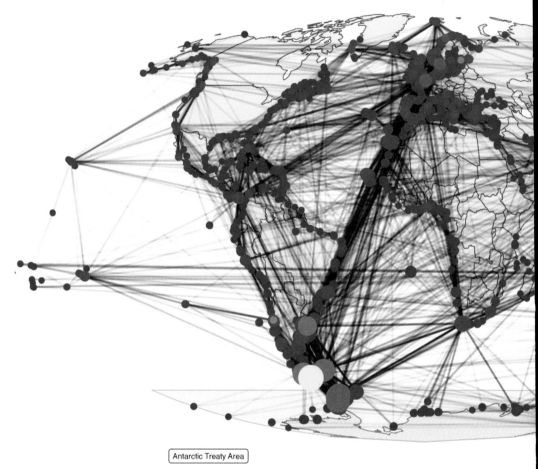

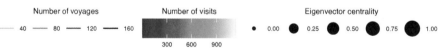

Antarctic Treaty Area

Number of voyages

40 —— 80 —— 120 —— 160

Number of visits

300   600   900

Eigenvector centrality

• 0.00   ● 0.25   ● 0.50   ● 0.75   ● 1.00

Map 16

Port to Port Traffic Network of All Ships that Visited Antarctica, 2014–2018

All heading south! Antarctica is getting busy

The first person to step foot on Antarctica was an English-born American seal hunter called John Davis. It was 7th February 1821, at a place now called the Davis Coast. In his log he wrote that the land was 'high and covered entirely in snow' and 'I think this Southern Land to be a continent'.

Antarctica is almost twice the size of Australia and far bigger than the USA or Europe. It has high mountains and broad valleys, and for millions of years it was left alone. After Davis's foray, few followed. The next visitor, the Norwegian, Henryk Bull, arrived some seventy-four years later.

Now look at it. The map shows all the ships that visited Antarctica between 2014 and 2018. They come from Europe, east Asia and South America, all headed for the last pristine place of any real size on the planet. Today about two hundred ships visit Antarctica each year. The majority make for the most accessible regions, shown by the yellow and blue dots. (Note: circle size represents the relative importance of ports within the network based on their connectivity to other highly ranked ports.) This is the Antarctic Peninsula: it's relatively temperate, easy to reach and you're likely to see whales, seals and plenty of penguins.

The growth of tourism is the number one reason for all the bustle. A lot of us have 'bucket lists' of destinations we dream of going to. The white, untainted wilderness of Antarctic is in many buckets. Tourist ships represent sixty-seven per cent of visits. In the peak season, December to February, more than 100,000 people visit Antarctica. Each year the number keeps rising (with the exception of pandemic year, 2020–21, which saw just 15 tourists). Future years will see this number rise: there are a lot of people in the world with a lot of buckets to fill.

The next biggest group of visiting ships are those making research trips. Fishing and supply vessels account for the remainder. The research vessels have the greatest number of connections across this map. It's this kind of connectivity that interests our map's creators, researchers from the British Antarctic Survey. They want to draw attention to the Antarctic's vulnerability to invasive species. The more boats and the more connectivity there is the more chances there are that unwelcome passengers, often clinging to the hulls of the ships, will be getting a free ride.

All parts of the world suffer from invasive species, disrupting ecosystems and endangering wildlife. It matters everywhere but it is especially important in the Antarctic because it has been isolated from other species for millions of years.

Gentoo penguins standing by a pile of trash from the Argentine refuge hut at Mikkelsen Harbor on Trinity Island, Antarctic peninsula

Antarctica is different. It is isolated in every way. There are no nations here. A variety of countries have made claims on slices of Antarctica, but they are not universally recognized. The United Kingdom, France, Australia, New Zealand and Norway form a cosy club in which each recognizes the others' claims, but other states do not. The most important legal tool that governs this remote land, the Antarctic Treaty of 1959, asserts that it is politically neutral. The Treaty also says that any nation that has signed the Treaty (fifty-six countries have now signed) has access to the whole region and that no country can use it for military purposes or as a dump for nuclear waste. More pressingly, although the Treaty includes a prohibition on mining, it runs out in 2048. An international conflict between protection and exploitation is brewing. Some countries are keen to begin drilling and digging.

In the southern summer about five thousand scientists and those that support their activities live in Antarctica. Tourists stay on board their boats, so their footprint is relatively light. It is the numerous research stations that define human life here and are responsible for the continent's villages as well as unfortunately the rubbish and debris that now adorn them. The largest base is McMurdo, an American outpost with about 1,250 residents, most of whom are support workers who keep 80 or so buildings ticking over. McMurdo has a fire station, shops and the continent's only ATM.

Antarctica is not a barren wasteland. It has few species of vegetation – liverworts, lichens and mosses – but the surrounding seas are full of animals, including globally important populations of whales and seals. On the land there are about twenty million penguins. All this life is protected from the outside world not only by how far away Antarctica is and how cold it is, but also by the way the currents of the Southern Ocean swirl round the continent, literally pushing away creatures, such as jellyfish and crabs, that venture towards it. Antarctica is a biodome, an exceptional and self-contained ecosystem that has evolved in isolation for the last fifteen to thirty million years.

Arlie McCarthy, one of the researchers at the British Antarctic Survey who helped create this map, explains how hands-on her work can be. She 'scraped hulls and pipes' she explains, 'freeing ships of their barnacles and sludge to find out what species are already being transported there and where in the world they come from'. Another of the British researchers, David Aldridge of the University of Cambridge, explains that 'non-native species are one of the biggest threats to Antarctica's biodiversity'.

Researchers have found a variety of non-native species now breeding in Antarctic waters, all introduced by shipping. They include crabs and mussels, and this matters because no similar creatures exist in these waters. Their presence could be game changer. Another group of scientists, based in Chile, have also found invasive mussels. Professor Leyla Cárdenas, from the Austral University of Chile, explains that mussels are 'excellent competitors for space' and could 'grow and rapidly dominate their surroundings'.

For millions of years Antarctica was isolated. No longer. Visitors of all kinds are making their way to the Earth's empty continent.

Railway map of People's Republic of China

Colored lines showing CRH and other high speed rail services

Last update: 2023-11-27

Map 17
Railway Map of China, 2023

China's high-speed revolution

This is a revolutionary map. That might surprise you. All it shows is a bunch of train lines; specifically, the highspeed railways across China. But these tracks upset centuries of stereotype and assumption. What this map tells us is that China is not 'developing', it's not 'catching up'; it's speeding by. About two-thirds of the world's high-speed rail track is in China. The network has grown rapidly and the plan is to extend it to 70,000 kilometres (43,500 miles) by 2035.

Our map reveals a dense network of red lines. Trains on these lines travel at speeds above 300 kilometres (186 miles) per hour. The yellow and blue lines are still high-speed but they offer a slower service, with trains running between 200 and 299 kilometres per hour (124 and 185 miles per hour).

This is also a map of population. Most people in China live in the east. The west has a lot of empty space. Nevertheless, rail is a key part of China's ambitions to trade globally. The yellow line that you can see on the map pushing deep into the west is a high-speed line that is part of the China's Belt and Road project, an on-going plan to extend China's power westwards, and one which is building transport links and changing economies across Asia, Africa and Europe.

Railways are a more efficient way of getting people and material around than roads or planes. Trains are cheaper and safer. The more people and material you have to move the more they make sense. It's true that, once in the air, planes go faster but over shorter distances trains get you to where you want to go quicker. Airports are nearly always a long way from town, they often involve queuing, invasive security, anxiety ('Please arrive at least two hours before your flight departs') and a lot of time spent checking in and out. Air travel also has a whopping carbon footprint. By comparison trains go straight to city centres, they involve far less hanging about and they are much greener. High-speed rail has eliminated short and mid-distance air transport between cities across China, Japan and Europe. It no longer makes sense to fly between Nanjing and Wuhan or Madrid and Barcelona. Everyone takes the train.

If you've had a look at Map 2, the Map of the Tracks of Yu (see page 12)', you might be thinking, what about the rivers? For thousands of years the waterways were China's transport arteries. They remain the most inexpensive way for transporting goods but they're too slow for people. The modern way of travel is not by road, or planes or waterways, it's by high-speed rail.

China has had railways since the 1870s and the network soon became extensive, but it was antiquated. Before the 1980s Chinese trains were largely powered by steam. That all changed, in large part because in 1978, premier Deng Xiaoping visited Japan and got very excited about the Shinkansen, the bullet train, the world's first high-speed locomotive. Japan was a world leader: its bullet trains had been zipping across the

Chinese high-speed trains

country since 1964. In 2007 China opened its first high-speed line and it has not looked back.

Well over two billion high-speed trips are taken in China every year. There are plenty of customers, but short-term commercial considerations take a back seat. This is true for every country that has high-speed rail. Without government investment, it doesn't happen. The scale of spend in China is colossal. The state-owned train operator's debts, in 2023, were reported as nearly $900 billion.

All the statistics are big numbers. Thousands of kilometres of high-speed rails have been built every year. The China Railway Construction Corporation employs upwards of 260,000 people. The company has transformed China and is building railways and roads across Asia and Africa too. Today three-quarters of Chinese cities with a population of 500,000 or more, have high-speed rail. China has also the longest high-speed rail line in the world, the one stretching 2,440 kilometres (1,516 miles) between Beijing and Hong Kong.

High-speed rail has taken off in a lot of places. Spain has Europe's most extensive network with a total of 3,600 kilometres (2,237 miles). Japan and France – both of which have been high-speed pioneers – have about 2,800 kilometres (1,740 miles). Many other nations are now investing heavily in high-speed, including Uzbekistan, Turkey, Morocco, Indonesia and Saudi Arabia.

At the other end of the spectrum there are countries with no or little high-speed rail. These include the world's poorest nations but also the USA and UK. There is no international agreement on what high-speed means, so the fact that a number of UK lines, up the west and east coast of Britain, allow top speeds of 201 kilometres per hour (125 miles per hour) might be used to argue that the UK already has a high-speed network. As a regular user of UK trains, 'high-speed' is not the first phrase that springs to mind. By most measures the UK has only about 100 kilometres (60 miles) of high-speed rail, a single track out of London to the Continent. The USA is also well behind the curve. Only Amtrak's Acela Express qualifies as high-speed.

Both the USA and UK have plans for more. The UK's HS2, a now reduced 210-kilometre (130-mile) city link, and California's High-Speed rail are due to open sometime towards or around 2030. Compared to what is happening elsewhere, they are small ventures, and they have both been beset with controversy. In both countries there is a squeamishness about tax-payers funding rail travel. Roads get the tax dollars that railways don't.

This railway map of China is a revolutionary map. It's a recalibration of world power. Some nations face forwards and some face back. China is going at high-speed. Many countries understand this and want to catch up. Others, such as the UK and USA, don't yet realize that they are in the slow lane.

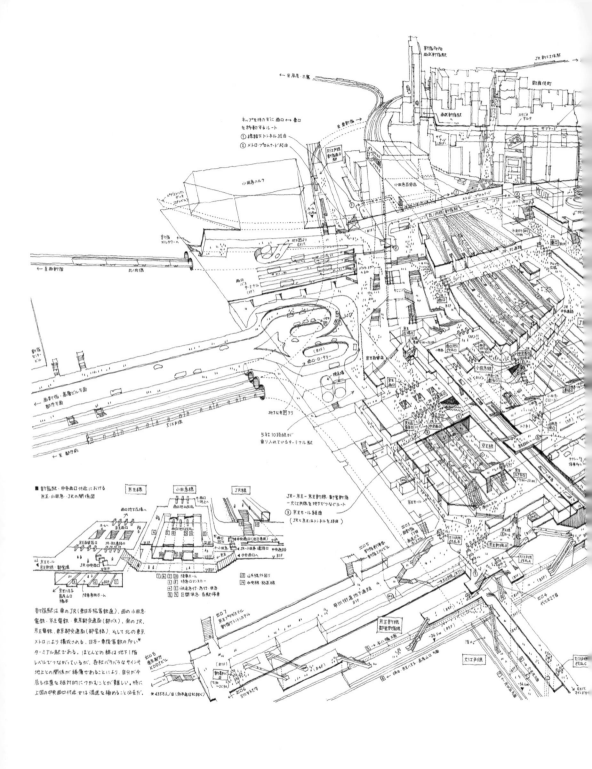

■ 新宿駅・中央西口付近における
京王・小田急・JRの関係図

新宿駅は東のJR（東日本旅客鉄道）、西の小田急電鉄、京王電鉄、東京都交通局（都バス）、南のJR、京王電鉄、東京都交通局（都電鉄）、そして北の東京メトロにより構成される、日本一乗降客数の多いターミナル駅である。ほとんどの線は地下1階レベルでつながっているが、各社バラバラなサインや地上との関係が稀薄であることにより、自分が今居る位置を相対的につかむことが難しい。特に上図の中央西口付近では混迷を極めること必至だ。 ＊435万人／日（乗車数値は別に開く）

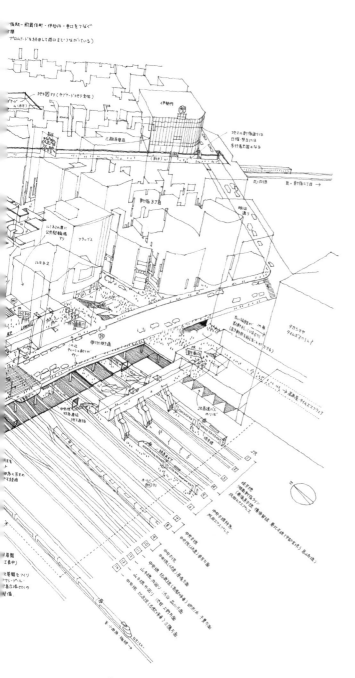

Map 18
Dismantling of Shinjuku Station, Tomoyuki Tanaka, 2005

A transparent map of a three-dimensional city

Japanese architect Tomoyuki Tanaka creates see-through maps of the city's multiple layers. He doesn't use computers but a pencil and a ballpoint pen. Simple tools, yet they open up a world that is so intricate and so strange that it looks like science fiction.

What we're looking at is a transport hub. With its two hundred exits and thirty-six platforms Shinjuku Station is the busiest train station in the world. It's used by about three-and-a-half million people a day. You can connect to the subway, inter-city, regional and airport lines, and every minute three thousand passengers alight or arrive. Each passenger must find their connection, or the way out, which is not always easy. There are many floors, stairs and escalators, and although the signage is clear, it can be daunting. Tanaka says that his drawing is 'a labyrinth' but not just that, it 'is almost like a living creature'.

This is not a map of chaos. Shinjuku Station works well: it is a perfectly timed operation, clean, efficient and it runs night and day. Yet it is immensely complicated and Tanaka's illustration invites us to get lost in its detail. It's all here: platform numbers, where the toilets are, the pathways and roads around Shinjuku that link it to the shopping malls that surround it. Is it a spider's web, a tapestry, a layer-cake? The similes don't fit because it's like nothing else. It's a landscape that overwhelms us. Yet if there is a 'living creature' in there, it is us.

Tanaka is a specialist in these kinds of X-ray cartographies. 'I want to show the relations between the internal and external spaces of the station', he explains. He's done other similar maps, almost as intricate, including of Tokyo's Shibuya Station. It takes him only a couple of weeks: a week spent on research, working out the layout of a place, then a week with his pencil and pen. The fact that this is hand-made object is important and telling. Japan has no shortage of computer power, but tradition and craft skills continue to be highly valued. Tanaka's hand-drawing expertise is admired: it's practical, skilful, yet conjures magic.

Shinjuku Station opened in 1885 as a stop on a new line operated by Nippon Railway, Japan's first private rail company. Old photographs show a low-rise building with a tiled roof and not a soul in sight. New tracks opened, traffic increased, and the station grew in complexity.

Most of us have experience of trying to navigate multi-level malls and interchanges. It has been obvious for some while that we need three-dimensional maps. Traditional maps are flat. They represent space as two-dimensional. That might work in a one-storey world but in today's cities, roads fly over each other, things are stacked and pathways, escalators and lifts are arranged in three dimensions. One usual solution is to draw multiple two-dimensional maps, one for each level. This works for a while but breaks down when levels start connecting and sprawling.

Our oldest reference point, the ground, is disappearing from view. You can see this is Tokyo but also in many other cities, such as Hong Kong. Hong Kong is estimated to have 60,356 lifts, though the most iconic element in its groundless infrastructure is its escalators. The

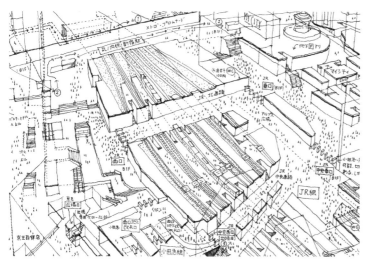

Detail of Shinjuku Station,
Tomoyuki Tanaka

Central-Mid-Levels escalators in Hong Kong comprise the longest outdoor covered escalator system in the world, and its direction of movement depends on whether it is ferrying commuters to or from work: from 6 a.m. to 10 a.m. the escalator moves downhill, and from 10.15 a.m. to 12 a.m., uphill.

There is a problem with three-dimensional maps. We can't understand them. Our cities have gained complexity but our brains, our spatial intelligence, hasn't. Tanaka's map is more art than a handy guide. Visitors to Shinjuku can make use of the much simpler three-dimensional maps provided by the station authorities. They are as simple as they can make them, but they are still hard to work out.

I got very lost in Shinjuku Station. That's no surprise because I didn't go there to catch a train but to hunt ghosts. It seems fitting that this 'living creature' should be a place of legend and mystery. It is sometimes called the Bermuda Triangle of Tokyo, and the story goes that some commuters never make it home. They take a wrong turn, then another, get flustered, run down the wrong stairs and end up in the wrong lift, until they find themselves quite alone in a quiet corridor, the soft boom of a distant underground train sounding somewhere far above them. They are never seen again. There are no verified cases, no roll-call of the disappeared. Tokyo is so big, so impersonal and machine-like that it is not hard to imagine that you could disappear inside it, that it could eat you up: that it contains phantom limbs, unnamed and hidden tunnels and lanes into which the unwary are drawn and where they are swallowed.

Shinjuku is also the home of coteries of ghosts. Not all of them are scary; one group of spectres has taken on the helpful task of saving people from suicide. They are said to push suicidal passengers away from the tracks to safety. These useful ghosts carry their own dark story; they were the victims of a secret mass suicide. They now drift about, making sure others don't suffer the same fate.

The modern city is intricate and, in Japan, efficient. Millions of people are ferried hither and thither, non-stop. Yet all these tangled lines combine to build a labyrinth that is beyond us, that can't quiet be grasped, and which it may not be possible to escape.

Map 19
The Global Flow of Internet Traffic, Barrett Lyon, OPTE Project, 2023

Big maps for big data

The internet links the world and this map shows how. It looks like an explosion in a firework factory, or a trippy flower blooming before dazzled eyes. It is made up of threads and patterns all dyed in vivid colours but what do they mean?

What we are seeing is a snapshot of the net from 2023, when the number of internet users in the world was more than five billion. It was created by Barrett Lyon, who as both artist and computer scientist is well placed to help out the rest of us, all those billions of scrollers and tappers who rely on the net but have little clue how it works.

This is a map of internet connectivity, the paths through which data flows from router to router. Lyon has arranged the net geographically, sorting it by its Regional Internet Registers. These are the organizations that administer and issue Internet Protocols (the IP code that all our devices rely on) and Autonomous System Numbers (sets of IP prefixes belonging to a network) within a defined region.

To most people this is unintelligible jargon, and I'm most people, but what it means is that this map is based on how permissions to run the internet are handed out. These permissions are given by real organizations, with buildings, squeaky chairs and flesh-and-blood people, and here each of those regional registers is given a colour. At the top where we have all those puff balls of green, is the European and Middle Eastern network. It's headquartered in Amsterdam, and there is a branch in Dubai. Blue represents the USA and Canada; red is the Asia-Pacific network, yellow the African network and pink the Latin America network. Lyon has also used two other colours: white for the backbone, or 'principal data routes' that connect networks and, somewhere in there but hard to see, brown for the network of the US military.

We can see instantly how the globe is portioned out and how different colours sprawl and collide. The closer to the centre of Lyon's map a network is, the more interconnected it is; and vice-versa.

The key take-away is how global internet use is today. Not many years ago this map would have shown a lot of activity in the West and much less elsewhere. Not so now: the internet is truly planetary. Is this a world escaping Western dominance? All the continents are present. Yet, given how dominant American corporations are in this sector and that much of the internet pushes English as its *lingua franca*, perhaps it is also a vision of a world in which direct Western (or at least American) dominance is no longer an issue because the West (or least America) is everywhere.

'The Internet is really big, very connected and extremely complex,' says Lyon. 'It's this whole world you can't see. That's the fun part of visualizing it.' Back in 2003 Lyon had a job as a 'penetration tester', someone who probes the vulnerabilities of computer systems. He developed software to do this automatically, mapping out clients' networks, but wondered if this could be scaled up. He floated an idea to friends: that he could map the entire net. 'They thought that was pretty funny', Lyon tells us, 'so they bet me fifty bucks I couldn't do it'.

Lyon's first version appeared in the early 2000s. Our map is a more recent version and also exists as an animation in which we can see the flowers bloom and fireworks explode as they grow over time and across every part of the world. This is a map of connectivity, but also an object of beauty, and it is little wonder that Lyon's work has been showcased in the Museum of Modern Art in New York. It's so pretty that it gives new life to the notion that the internet is a kind of magic; an ethereal force of ebbs and flows.

Although we use airy concepts such as 'the Cloud' to talk about data storage, real, physical cables do the work of trafficking our data. It's useful to pair Lyon's magisterial creation with another, also reproduced below, which plots the most important of these cables, which lie under the sea. There aren't that many, though the number keeps growing: about five hundred cable systems carry almost all the world's transoceanic data. All the continents, apart from Antarctica, are now connected by submarine cables. We can also see the density of connections in east and south-east Asia and in the north Atlantic. Less populated areas have fewer cables, as can be seen from the lack of connectivity to and, hence, invisibility of, Western Australia, tucked in the bottom right of the map.

This is a new world but its history is worth telling. In 1844 Samuel Morse sent the first telegraph message. It travelled from Washington, D.C., to Baltimore, and read 'What hath God wrought?' The first transatlantic cable arrived in 1866; a cable from India to Yemen came in 1870. Recent decades have witnessed a revolution in the industry in terms of both the types and amount of data that can be sent. Old style cables were just point to point. These days submarine branching units allow a single cable to connect with multiple destinations. The first transatlantic optical fibre cable went live in 1988.

Undersea cables offer a much better service than satellites. Not only are the speeds much faster but cables can communicate tens of terabits of data per second, vastly more than any satellite. These cables use light to encode information and can send data at speeds approaching the speed of light. They are unaffected by weather and apart from the odd shark gnawing them, have proved to be resilient. Only when capacity undersea is reached do telecommunications companies switch to satellites, as a second-best stand-by. The internet is not a magic show. It is everywhere and connects us all but also it is somewhere, flashed along cables that scroll the murky seabed.

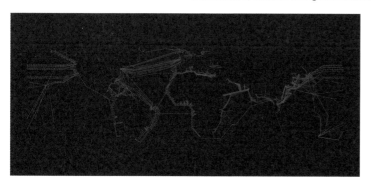

Submarine internet cables

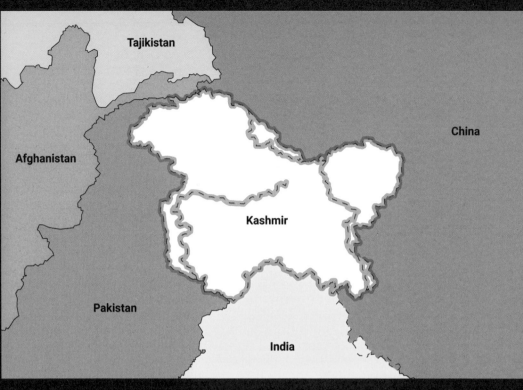

Map 20

Kashmir for Google users in India versus Kashmir for Google users outside India, 2024

Google's chameleon maps

There's no getting away from Google. It dominates our access to knowledge and our maps. It doesn't have total control, thanks to other providers such as Apple Maps. But these two companies do more than take the lion's share of the online map market, they are the market. Where this isn't true, such as in Russia and China, it's because the state has stepped in to carve a space for local alternatives.

Google is the prism through which we see the world. But this prism can be turned. One way of doing that when using Google Maps is by changing your 'user country'. I'm in the UK so my default 'user country' is set to the UK. But I can alter that and, when I do, something interesting happens.

Google tailors its output. If it thinks you are in Russia, then it will show that Russia includes Crimea (with a hard border). Crimea is indeed under Russian control. That's because Russia invaded Crimea in 2014. What if we make Google Maps think we're in Ukraine? We see something different: Russia's border drops south and now Crimea is part of Ukraine. Elsewhere, the border is shown as uncertain with a dotted line.

Just as contentious is the way that, for users in Turkey, Google Maps names the Turkish Republic of Northern Cyprus. This is the northern part of the island of Cyprus and was created by force when Turkish troops landed there in 1974. It's not a label or a country recognized by any other nation in the world. Just Turkey. Users outside Turkey see some dotted lines, of uncertain meaning, scissored across the island but zero sign of the Turkish Republic of Northern Cyprus.

Place names matter and a lot of places have contested names. Sometimes you have to change the language option as well your location to see how Google tries to sail these turbulent waters. Only when you switch to Chinese do you see that users in Hong Kong (there is no option for setting China as your location) see Yongshu Island in the South China Sea. This is the Chinese Government's name for a once pristine reef now not just claimed but covered in concrete and turned into a Chinese military outpost and airstrip. If you keep to English, you see Fiery Cross Reef, its internationally recognized name. The English name carries a hint: that, despite occupying the island, neither it nor the many other Chinese military island bases that freckle these waters, are an internationally recognized part of China.

For most of us the sea between Japan and Korea is called the Sea of Japan. But this is one topic on which North and South Korea are of one mind: it isn't the Sea of Japan it's the East Sea. So that's the name, for users in South Korea, it goes by in Google Maps. I guess that would be true for North Koreans too – if they had access to foreign search engines or, indeed, the internet, which most don't.

A final example. Over in the South Atlantic, Argentinians see Islas Malvinas far off their eastern coast. It's a cluster of islands also claimed and, more to the point, occupied by Britain. Britons see the same place but with a different name, the Falkland Islands.

Crimea for Google user in Russia

Crimea for Google users outside Russia

Google's chameleon maps adapt, bend and blend into the territory they seek to serve. Why does this matter? It's not because different maps show different things. Maps have always differed. They probably differ less now than at any time in the past, as globalization and the internet have drawn us closer together. It's not that nations disagree about borders and names that is at stake here but that one organization, a private company based in the Googleplex campus, Mountain View, California, is in a position to depict, communicate and, as it sees fit, accommodate these differences. That's a lot of power.

The internet grants users access, nothing more. Google keeps a tight hold of its copyright. Every Google Map you see is company property. Users – you and me – are granted *permission* to see them. This is why we had to draw our own version of Google's maps of Crimea and Kashmir for this book.

Kashmir is a country of snow-white mountains and wide green valleys that should be full of tourists but is full of soldiers. It sits at the top of India and the eastern side of Pakistan. In both countries, Google's share on the mobile search engine market is almost a hundred per cent. These maps matter. The bottom one shows us what Google Map users in India see. It portrays the whole of Kashmir, not just the bit that is actually administered by India, as if it were under Indian control. The top map is what you see if you're a Google Map user living in Pakistan, or anywhere else, and is more complex. It has a lot of dotted lines, which indicate disputed lines of control, in this case not just between Pakistan and India but also China, which claims and controls the dotted zones on the eastern side of the map. One of those dotted lines – the one that marches out across the middle of the territory – is so uncertain that it appears to give up. It stops, paralyzed by worry, in a high, cold nowhere, or more precisely in a frozen slope in the Himalayas, which separate Kashmir from China.

What neither map shows us is that plenty of people in Kashmir don't want to be in India, Pakistan or China. They want to be in a country called Kashmir. It's an unlikely prospect, given how much national pride Kashmir's neighbours pour their way. Kashmir is one of the most militarized places on Earth. Hundreds of thousands of troops keep it in a vice-like grip and there are continuous skirmishes and bouts of violence. Thousands of civilians have been killed and, currently, there is little sign of peace. In 2019, India reinforced its claim on the parts of Kashmir that are under its control (which is most of the southern half) by getting rid of the political autonomy the region once enjoyed and governing it directly from New Delhi.

Google Maps stands aloof. The view from Mountain View is Olympian. The well-paid techies at Googleplex gaze down and try to puzzle out the foibles and prejudices of their users and give them what they want. Google is a world maker. Like a Greek god, it ponders the realm of the mortals, then decides what is best.

Map 21
Walkability for Women in New York City, 2021

A women–friendly city is something to be

Women and men have a different relationship to the geography of fear. It's true young men are most likely to be attacked on the street, a fact often rolled out in conversations about women's safety, with the non-too-subtle hint that women's concerns are driven by emotion not reason. Yet young men, and men in general, are even more likely to be perpetrators of street violence. If you're in a fight you stand a good chance of getting hit.

Men are far less likely to be the victim of the kind of violence both men and women fear the most, sexual violence. Let's be even more honest: fear isn't about crime statistics. Being stared at, being followed, being shoved, being got too close to, being whistled at and jeered at ... the number of ways that men make women feel unsafe are legion. Vanishingly few make it onto police records and into official statistics. This kind of harassment matters: it combines and congeals to make public space feel like hostile space.

What we are looking at in this map is New York colour-coded according to a Walkability for Women Index. Places in green are neighbourhoods that feel and are safer. It is along these streets that women might get into a routine of going for a walk or a jog, places where the trip from work to home, from a bar to a friend's apartment, will be less anxious.

Most of the neighbourhoods around Central Park in Manhattan are inked in deep green (though they start shading yellow from the park's top end, where Harlem starts). These leaf-coloured areas are safer but they aren't Eden. Even in these safer parts of town, women will still be planning how they get from A to B in a way that men don't. Women will still be making sure they have company when walking home after dark, arranging to go running with a group rather than taking the risk of being alone. It's common sense, second nature, but it's not nature that is at work but the way city space is coloured by male violence.

The green zones are not paradise but they a lot better than the areas in orange and red. These are neighbourhoods that don't pass the walkability test. It's plain to see how Manhattan and the richer parts of Brooklyn (which are both to be found in the middle of the map) hog the green but that there is a fat ring of other colours around these upmarket neighbourhoods, where walkability plummets. It's a thick band that reaches across Staten Island (at the bottom left of our map), through lower Brooklyn and sprawls over most of Queens (on the east side of New York) before curling up into the Bronx (a borough in the north of the city).

Bad places for walking are usually poorer places. Wealthier areas are cleaner, better maintained, have more bustle and feel cared for. In most cities, you only have to walk a few minutes from the centre to find streets that feel and look rougher: things aren't mended, stuff is broken, rubbish is piled up and graffiti covers the walls.

Woman running at night

Somewhere like Far Rockaway, a patch of red tucked in the Rockaway Peninsula, an exclave of New York that juts out from Long Island and is found at the bottom right of our map. It's part of the borough of Queens, which generally does badly in this survey. Far Rockaway is marred by low incomes, high crime and, according to the researchers who created this map, is one of the least walkable places in New York for women (and it's probably not so great for men either).

The relationship between money and walkability is strong but it's not an iron bond. Per capita income on Staten Island is second only to Manhattan – and more than a third higher than the Bronx – but the island glows with oranges and reds.

How can you make a map of walkability? This one was drawn in 2021 by a team of four environmental psychologists, three men and one woman. Their work was helped by the fact that the US already has a National Walkability Index. They used this and then looked at a number of other factors, including women's perceptions of places that feel safe. The team also made use of what they called 'vitality of the social context or attractiveness' and 'comfort of the road infrastructure in terms of sidewalks, public space, and green areas'. They may sound long-winded and a bit woolly, but the team pinned each down to observable, real things, such as sidewalk width, and number of road crossings, cafes and restaurants. They also included recorded crimes, as reported to the New York City Police Department in relation to harassment, rape, obscenity and abuse.

This map matters: it's not based on speculation but real data. One thing it does miss is time of day. It offers a generalization across the whole day. It's worth making an obvious point: when it comes to walkability, night and day are not the same. It's at night that many streets, from New York to New Delhi, turn toxic.

Maps are not just illustrations, pretty pictures, they pose questions and push for change. Why shouldn't women be able to walk out of their front doors without fear? Why should it be so much harder to do this in Far Rockaway than mid-town Manhattan? Cutting off citizens from public space has serious consequences. This isn't just a map about violence, fear and crime. It's also a map about health. Walking is by far the easiest and most important daily routine that keeps people in good physical shape. If the street and parks are no-go zones then those who can afford it will drive off to gyms but, overall, people will move about less. They will get less healthy, they will feel more confined, and they will bump into fewer people, have fewer friends. Maps of walkability are also maps of well-being, sociability and connection.

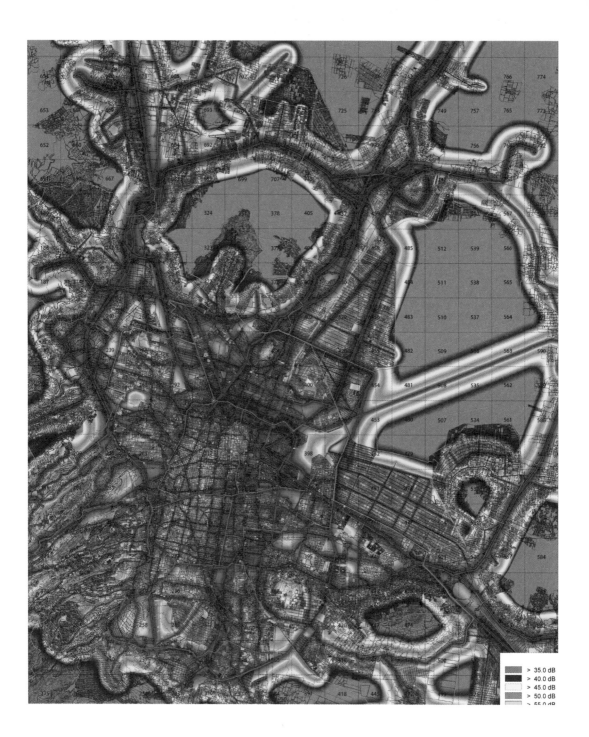

	> 35.0 dB
	> 40.0 dB
	> 45.0 dB
	> 50.0 dB
	> 55.0 dB

Map 22

Noise Map of Mexico City, Universidad Autónoma Metropolitana, 2011

Overwhelmed by the noise of cars and trucks

This is Mexico City but it could be any city. Over the past one hundred years we have built a noisy world. Many of us endure levels of disturbance, day and night, that would have appalled our ancestors. Like most bad habits, it crept up on us, little by little, decibel by decibel.

Many of us are still pretending that noise pollution is a minor annoyance, like having a shirt collar that rubs, or a jagged fingernail. A new generation of acoustic mappers, however, are showing the truth. Noise wrecks lives, it makes people ill and it's everywhere.

With a population of twenty-two million, Mexico City is the largest Spanish-speaking city in the world. Anyone who's visited knows that one of the most dangerous things you can do in Mexico City is to try and cross the road. Despite its millions of people, the car – and there are more than five million of them in this city – is king. Tranquil valleys and hills surround the city, especially in the northern suburbs, picked out in green on this map, and there are big parks in the east. They are calm retreats but for the most part, it is road traffic that swathes this map in red and purple, the colours that show high noise levels. It looks like a diseased road atlas.

Mexico City has a charming and historic old city at its centre and it's here we see a small patch of lighter colours. Around the famous central square, with its vast cathedral, Aztec archaeological sites and constant hubbub of people and demonstrations, there is a no-car policy, enforced for most of the day. These kinds of bans were bought in, as they have been in other cities, to improve air quality. According to the World Bank, air pollution kills nearly 33,000 Mexicans every year. The slight reduction of noise pollution was a side-effect.

One of the twenty-first century's new professions is 'acoustic mapper', and within that community this is a famous map. In 2011, a group of civic organizations in Mexico City got together to create it and it became a founding statement in the field of environmental acoustics. It's not a fine-grain vision but a city-wide wake up call. It goes for the jugular. Any map that wants to shift debate on an issue needs to be easily and immediately understood. This one is: it reveals a sick city. Those feverish colours, those fat pulsing veins; it's a cartographic migraine. The message is simple: noise matters!

The creation of this map involved the analysis of around 2,200 kilometres (1,370 miles) of roads. Building works, noisy bars and noisy neighbours create their own noise burden, which can be especially terrible because they are erratic, unpredictable and can carry a sense of threat. The sound of a smashed bottle and/or a scream is upsetting, irrespective of number of decibels. Nevertheless, cars and trucks are in a league of their own. And scooters. Researchers at Bruitparif, a pioneering group of scientists who monitor noise in Paris, found that a single unmuffled scooter crossing Paris at night can wake as many as 10,000 people.

The high noise ranges shown on the map go from 75 to 85 decibels, which is about the same as that made by a blender, lawn mower or subway train. If you are exposed to this level for eight or so hours a day,

Downtown Mexico City, with Torre Latinoamericana in the foreground

it will lead to hearing loss. Today a lot of city people have impaired hearing. A study by researchers at the University of Michigan discovered that that eight out of ten New Yorkers suffer hearing loss.

There is a clear correlation between wealth and exposure to noise pollution. Street vendors and other people who live hard by the road in Mexico City get the worst of it. Rich people live behind gates in quieter neighbourhoods; or high up, away from the grind of gears and wheels on tarmac behind double- or triple-glazed windows.

Some readers may still be thinking, it's just noise, how bad can it be? It's not one of the ills we are taught about at school. Smoking, alcohol, drugs, air and water pollution, these we know about. But noise?

One way government researchers in the UK sought to answer this question was by using an internationally recognized metric called Disability Adjusted Life Years or DALY. One DALY represents the loss of one year of good health. They found that in 2018, around 100,000 DALYs were lost in England due to road traffic noise. A further 13,000 were lost from railway noise and 17,000 from aircraft noise. Most of these losses were due to chronic annoyance and sleep disturbance, followed by stroke, ischemic heart disease and diabetes. The European Environmental Agency reports that, across Europe, noise caused thousands of premature deaths and that twenty-two million people suffer 'chronic high annoyance' and six and a half million people suffer 'chronic high sleep disturbance'. I suspect these figures are the tip of a very large iceberg. Noise disturbs us, at a deep level. It upsets us and makes life unbearable.

Who doesn't yearn for peace and quiet? We know it's important, but we're confused about what to do about noise. This bewilderment is reflected in the law. In most places there are either no laws that cover it or the laws that do exist don't work. People throw up their hands, as if it were too complicated to figure out, as if stopping building huge roads through residential neighbourhoods for example was impossible. The changes required are not rocket science, nor are they out of reach. It's one of the easiest ailments to fix. If we think of peace and quiet as a human right that would help. There is no good reason why, one day, this noise map of Mexico City should not be bathed in green.

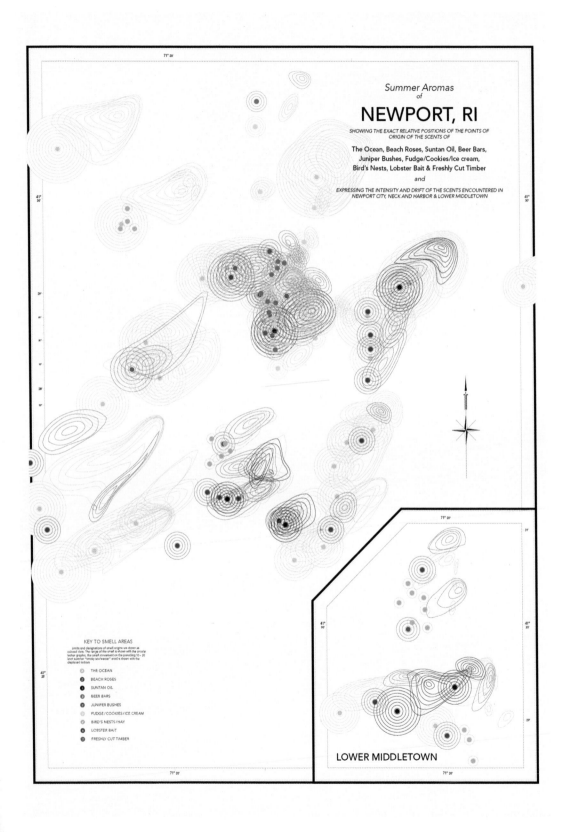

Summer Aromas
of

NEWPORT, RI

SHOWING THE EXACT RELATIVE POSITIONS OF THE POINTS OF
ORIGIN OF THE SCENTS OF

**The Ocean, Beach Roses, Suntan Oil, Beer Bars,
Juniper Bushes, Fudge/Cookies/Ice cream,
Bird's Nests, Lobster Bait & Freshly Cut Timber**

and

EXPRESSING THE INTENSITY AND DRIFT OF THE SCENTS ENCOUNTERED IN
NEWPORT CITY, NECK AND HARBOR & LOWER MIDDLETOWN

KEY TO SMELL AREAS

Limits and designations of smell origins are shown as
colored dots. The range of the smell is shown with the circular
isobar graphs, the smell movement on the prevailing 10 – 20
knot summer "smoky southwester" wind is shown with the
displaced isobars.

- THE OCEAN
- BEACH ROSES
- SUNTAN OIL
- BEER BARS
- JUNIPER BUSHES
- FUDGE/COOKIES/ICE CREAM
- BIRD'S NESTS/HAY
- LOBSTER BAIT
- FRESHLY CUT TIMBER

LOWER MIDDLETOWN

Map 23
'Summer Aromas of Newport, Rhode Island', Kate McLean, 2012
Mapping the smells of a seaside vacation

Most animals use their noses to guide them. The twitching snouts of mice, dogs and pigs are navigational tools. The human nose is equally prominent on our faces, but smell remains our most disregarded sense: sights and sounds matter to us in a way that aroma does not. The word 'odour', along with 'smelly', is used to describe low-status things to be avoided. But smell is everywhere and, as other animals know, it packs a lot of information.

This is what designer and cartographer Kate McLean has been showing in a series of rather lovely 'smell maps' of cities in Europe and America. This one is of Newport, the pretty harbour-side town on Aquidneck Island, Rhode Island, USA. Newport draws the crowds for its yachting life, grand mansions, flower gardens and seafood.

All those swirling lines makes it look like a contour map of hills and valleys, but McLean's map conjures a far more transitory landscape. It charts summer aromas, the whiffs of a carefree seaside holiday. The different colours represent different smells: expressing, says McLean, 'the intensity and drift of the scents encountered'. It's an idiosyncratic list: the pale blue is the smell of the ocean; red: roses; dark blue: suntan oil; orange: beer bars; mid-blue: juniper bushes; green: what McLean classifies as 'fudge/cookies/ice cream'; mauve: 'bird's nest/hay'; dark red: 'lobster bait' and cobalt blue is 'freshly cut timber'.

Just like with contour lines, the closer the lines get the more intense the experience. The way they loop and extend shows the drift and range of any particular smell. You can tell this is a coastal town from the way the tang of the ocean is so prominent, especially harbour-side, but it folds into other scents, such as the sweet red smell of roses. The busy tourist hive at the map's centre pulses and collides with more mouth-watering traces, the hoppy stench of beer and the sugary rush of 'fudge/ cookies/ice cream'. Further out smells come and go, meeting and connecting the woody, spiciness of juniper bushes with the hot reek of suntan oil.

McLean tells us that her map was assembled from talking and listening to residents and visitors

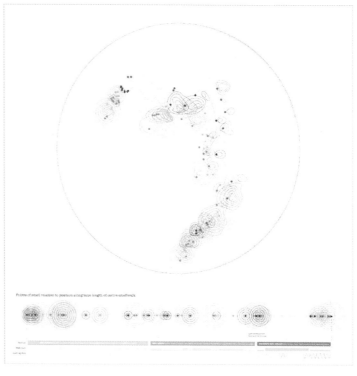

'A Winter Smellwalk in Kyiv', Kate McLean, 2016

as well as 'through smellwalks and smell bike rides and the nose of the artist'. Elsewhere she writes that 'memories triggered by smell are more emotional than those triggered by sounds, pictures or words'. Somehow this most disregarded of the senses, can transport us, instantly. McLean's map is an evocation and a provocation: it challenges us to value smell.

McLean's smell maps are more art than science and are sold as attractive wall posters. Yet they open up important questions about our relationship to smell. The widespread assumption that, compared to that of other animals, the human sense of smell is very poor turns out to be untrue. In 2006, scientists at the University of California did a unique experiment comparing the smelling power of dogs and humans. Obviously, dogs' is better? Yes but also no. The researchers covered twine in chocolate and dragged it across a grassy field then asked blindfolded volunteers to go down on all fours and track the scent. No surprises. To start with. The crawling and bumping volunteers weren't as good as dogs. But, with practice, they got better, a lot better. It appears humans, like other critters, *can* follow a scent trail. In fact, there are certain smells – apparently the odour of bananas, flowers and blood are among them – that humans are better at detecting than other animals.

Increasingly scientists are arguing that the idea that humans have a weak sense of smell is a myth. It turns out that the number of neurons found in the olfactory bulbs of mammal brains – including in humans – is pretty consistent. Humans don't smell badly. They simply think they do. Just as intriguing is the fact that women have slightly more of these neurons than men. Women literally smell better.

McLean's maps are the cartographic and artistic front of a wider push to reassess our sense of smell. She explains her map–making process by saying it 'starts with me manually transcribing the smellwalker data into a database and looking for patterns and commonalities'. She then chooses 'the most common smell in a given neighbourhood'. At other times she maps 'individual perceptions in specific streets to illustrate the differences'. This was the approach she took to create 'A Winter Smellwalk in Kyiv', which we reproduce here. It plots seven individual trails showing 'smell sources and projected smell shapes in the light, westerly winter wind'. McLean explains that 'each of the seven smellwalkers is represented by a colour indicative to them of the Kyiv smellscape'. The smells her smellwalkers record included: 'cheap food and coffee'; 'humidity fresh dampness' and 'smell of cheap plastic souvenirs'. Along the bottom of the map can be found a line depicting her walkers' stopping points, which McLean calls 'cumulative pulse points of smell detection'.

These maps touch on powerful truths: the uniquely evocative nature of smell, the way it conjures memory and how landscapes are swathed in scent. In some ways we are only starting to understand our relationship with smell. The vocabulary of smell is evocative and metaphorical but, somehow, never quite adequate. Perhaps smell is beyond words. In researching her maps McLean has heard some very odd descriptions of smell. People have described to her odours as allusive as the 'smell of shattered dreams', the 'smell of local government', and 'moments of joy', which – oddly but somehow understandably – turned out to refer to the smell of cigarette butts.

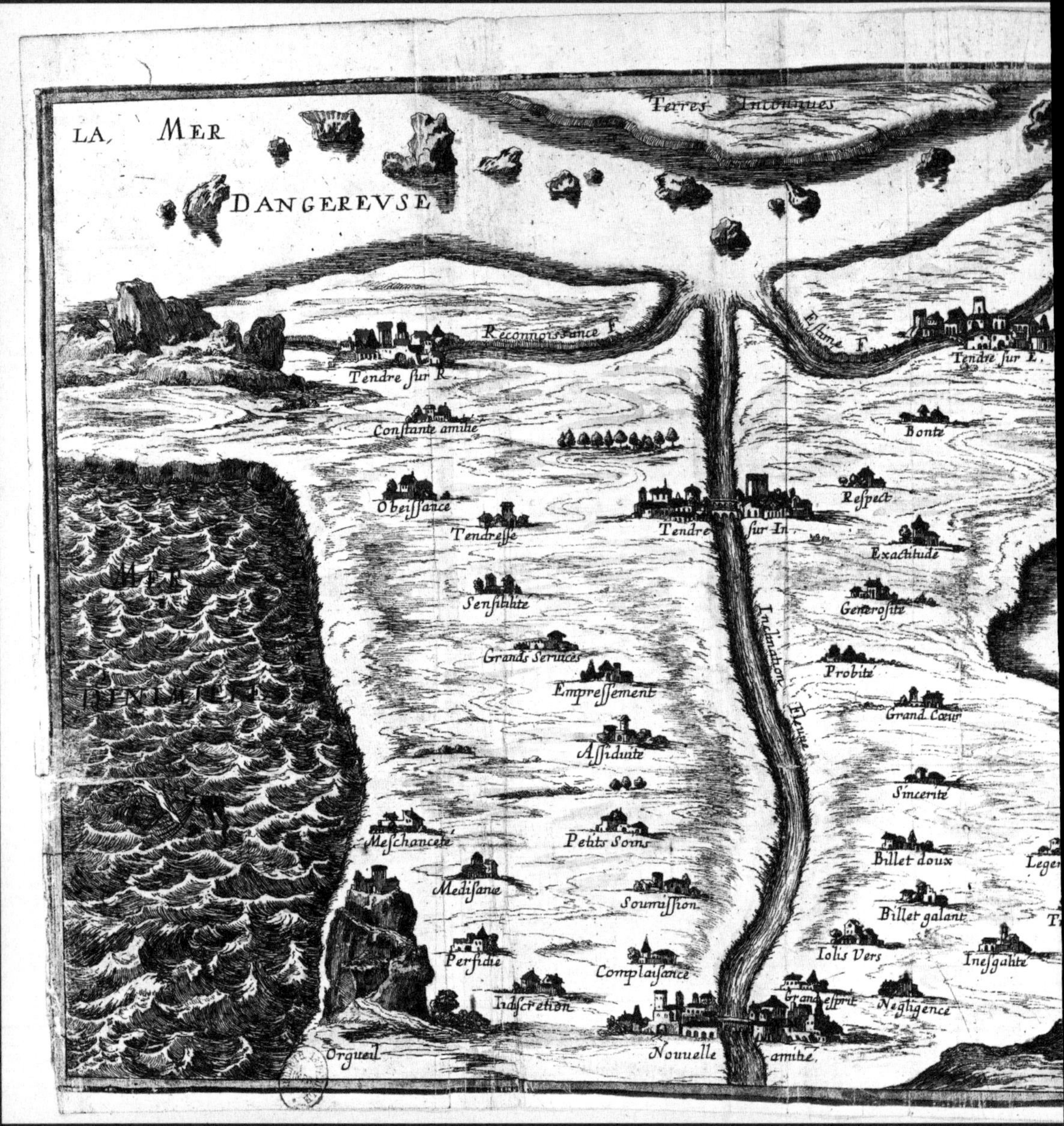

LA MER

DANGEREVSE

Terres Inconnues

Reconnoissance F.

Tendre sur R.

Estime F.

Tendre sur E.

Constante amitie

Bonté

Obeissance

Respect

Tendresse

Tendre sur In

Exactitude

Sensibilite

Generosité

Grands Seruices

Probité

Empressement

Grand Cœur

Assiduité

Sincerité

Meschanceté

Petits Soins

Billet doux

Leger

Medisance

Billet galant

Soumission

Perfidie

Iolis Vers

Inesgalite

Complaisance

grand esprit

Negligence

Indiscretion

Orgueil

Nouuelle amitié

Map 24
Carte de Tendre
'Map of Tenderness', 1653

A game of love

Love has destinations and it has starting points. The human heart can be mapped and for upper-class Parisians in the seventeenth century there was no better guide than the *Carte de Tendre* (Map of Tenderness). The novelist, high society host and toast of Paris, Madeleine de Scudéry, appears to have had it produced it for the amusement of her well-heeled guests in 1653.

An invitation to Scudéry's 'Saturday Society' was coveted but must have occasioned a few butterflies. Her salon, and this game in particular, required a lively wit. Think of it as a board game in which players role-play a would-be lover. To begin, we are asked to imagine ourselves as one of the gallants grouped on that hummock at the bottom. They are deep in conversation, working out the best pathway through the landscape that ranges in front of them. Nouvelle amitié, the town of 'New Friendship', at the bottom centre of our map, will be their starting point. So they have a woman's new friendship, now they want to win her love. The challenge is to spin a story, one that will lead her to one of the three towns in the far distance. Each offers a different type of love: Tendre-sur-Estime, (top right) Tendre-sur-Reconnaissance (top left) and Tendre-sur-Inclination (just above the centre) (Respect, Recognition and Inclination).

There are no dice, no rules and no limits. To play you need to tell a compelling and pleasing tale which, hopefully, will provoke twinkling assent and laughter from the finely dressed folk that have gathered around you.

All very refined. Yes and no. This is a game of wit but also of sex. Any possible resemblance between the River of Inclination, which cuts through the middle of the map, and a vulva, or the spreading landscape above it and pubic hair, would not have been wasted on the mirthful players.

You may notice that the quickest route to love is via that river. But try not to rush to conclusions. The smooth waters of 'inclination' – the sympathy and empathy between a man and woman – may be direct but if that's your single idea and all you can come up with, you risk stifled yawns.

Who can do better? Who can amuse us? All those little villages are conversational opportunities. I can see Petit Soins (Pamperings), Tendresse (Tenderness) and Obéissance (Obedience). To win a lady's heart and arrive at the town of Recognition we must surely show all three. Here's another pathway: if we go by way of Generosité (Generosity) and Billet doux (Love Letter) we arrive at Esteem. Let's not to forget to say something clever about the many dangers that lurk. There is the castle of Orgueil (Pride), atop a lonely mountain, bottom left, and hemmed around with the dismal villages of Meschanceté (Meanness), Indiscretion (Indiscretion), Medisance (Disparagement) and Perfidie (Betrayal). And I wonder what could lurk in La mer dangereuse (The Dangerous Sea), or the Mer d'Inimitié (Sea of Enmity') and how to describe the Terres inconnues (Unknown Lands) without becoming salacious?

Not funny? Wondering if there are better ways to spend a Saturday? Don't blame the host. The game was as good as the players and Madeleine de Scudéry was creating something new. She had invented the first role-playing game. It could also be claimed as a distant ancestor of the numerous map-based landscapes that dominate modern board and video games. Dungeons and Dragons or Assassin's Creed with saucy banter? Perhaps not, but certainly a fascinating and early example of a social game.

Madeleine de Scudéry must have been an impressive presence at her Saturday salons. She was more than just a host for the idle rich. She boldly asserted the right of women to have their voice heard. Her Les Femmes Illustres was published in 1642 and is one the earliest feminist statements in any language. In it she argues for female education and economic independence and against the orthodoxy that culture, politics, discussion and debate, was only for men. These were pioneering ideas and they give us another window into the Saturday Society. Amid the mirth and sexual dalliance something else can be heard: a fight for basic human rights.

Writing under her own name or the pseudonym of Sapho, Scudéry wrote fantastically long novels centred on the possibility of human affection and attraction. This map has another life as a frontispiece for one of them, a novel named after its romantic heroine, one of the very first in literature, Clélie. It appeared between 1654 and 1661 and sprawls over ten volumes. Scudéry's other novels were of similar length and though once widely translated, they are now little read.

Reproduced overleaf is another map of love that's hard going, but in a different way. It comes from a pair of heart-shaped maps published in the USA in the 1830s. A Map of The Fortified Country of Man's Heart is walled in with ramparts. Its outer crenulation is called 'Dread of Matrimony' and at its centre is the 'Citadel of Self Love'. The lands of a man's heart include 'Love of Power', 'Love of Money', 'Love of Ease' and the 'Land of Economy'. The 'Land of Romance' occupies but a small corner of Man's Heart. The map contrasts with its pair, A Map of The Open Country of Woman's Heart which contains the lands of 'Coquetry', 'Love of Dress' and 'Sentiment'. A wide watery channel leading to its centre gushes in from the 'Sea of Wealth'. Both maps are satires and anonymous but said to be the work of 'a Lady'.

Scudéry's map and this much later cartography of the human heart come from periods of history when male dominance went largely unquestioned. They are equally stuffed with stereotypes about what men and women are like. The Country of Man's Heart still hits home, for these clichés and stereotypes are with us today. But I think Scudéry's map of love contains an alternative history. It was at the beginning of so many things, of social board games but also of the assertion of social life as more than simply a man's game. While the later, nineteenth century, map highlights yet fortifies gender clichés; the other, much earlier map hints at multiple pathways and destinations.

Overleaf: 'A Map of The Open Country of Woman's Heart' (left),
and 'A Map of The Fortified Country of Man's Heart' (right), 1830s

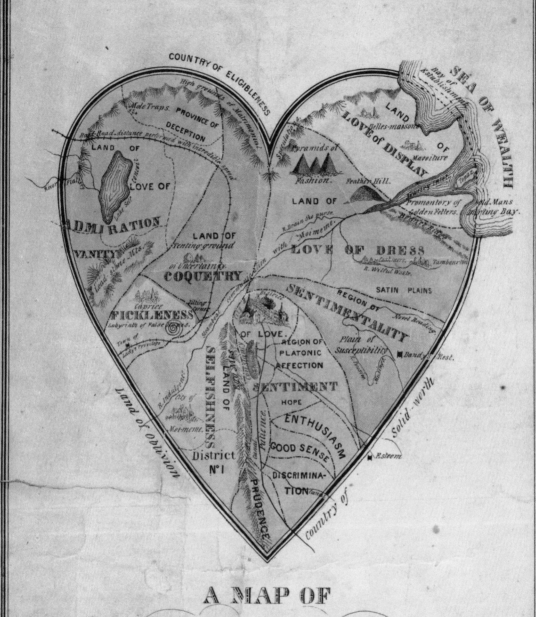

A MAP OF

THE OPEN COUNTRY OF

WOMAN'S HEART,

EXHIBITING ITS INTERNAL COMMUNICATIONS AND THE

FACILITIES AND DANGERS TO TRAVELLERS THEREIN.

By a Lady.

Kelloggs & Thayer, 144 Fulton St. N.Y.   E.B.& E.C. Kellogg 136 Main St Hartford Conn.   D. Needham, 223 Main St Buffalo.

Entered according to Act of Congress in the year 1846 by Kelloggs & Thayer in the Clerks office of the District Court of the Southern District of N.Y.

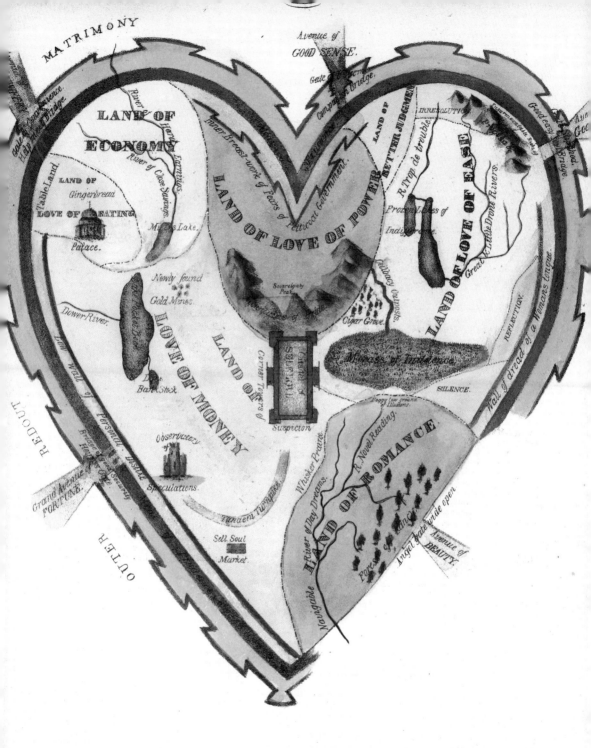

THE FORTIFIED COUNTRY OF
MAN'S HEART,

Exhibiting its defences, and modes of exposure to attack.

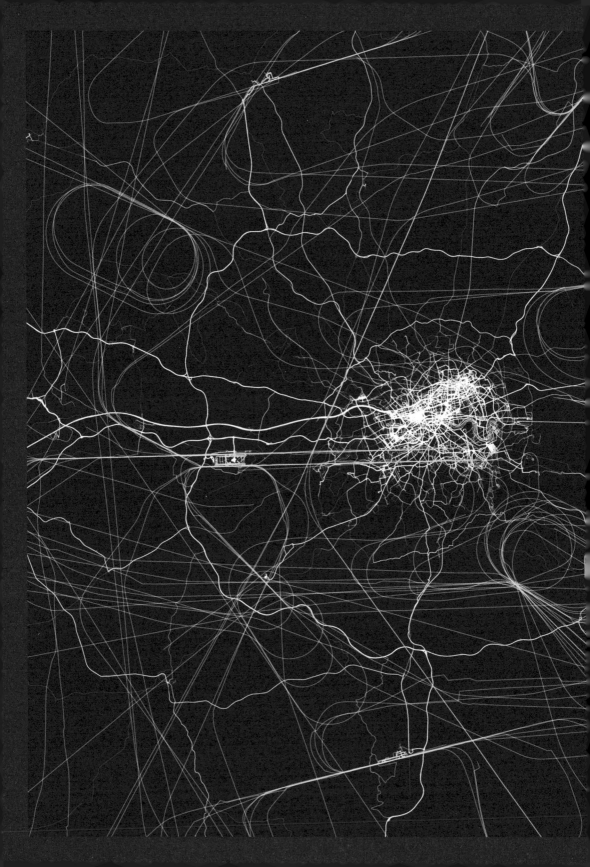

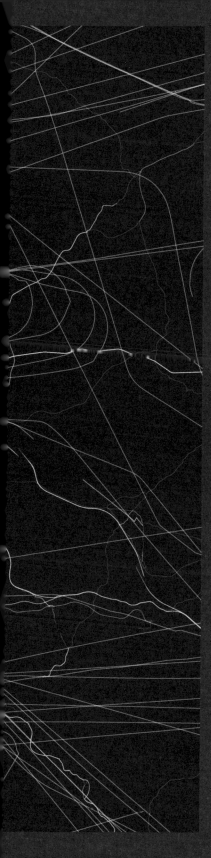

Map 25
My Ghost,
Jeremy Wood, 2009

Wandering ghosts: the art of GPS

Maps fascinate contemporary artists. They are drawn to their power to conjure landscape and movement. While a painting, a film or a book can represent a place, a map does something more: it shows us the whole, it guides us and it invites us in. Maps can be science and magic.

You don't have to know a thing about it to enjoy this map. Golden and white wires spiral, leap and zip about in an inky void. There is a lot of restlessness here but what we're seeing are traces from the past, like a sparkler's trail effect, still imprinted on our eyes after the light has passed on.

It's called *My Ghost* and was created by the digital artist Jeremy Wood. For fifteen years everywhere Jeremy went he took with him a GPS tracker (GPS means Global Positioning System and refers to a network of satellites that locates enabled devices which, today, includes most phones). He used it to record his movements across London. GPS was Wood's brush or pen; he calls it 'a tool that uses your whole body to draw with'. What he ended up with, he says, 'is a form of personal cartography that documents my life'.

The yellow lines are Wood's travels in planes, looping in holding patterns above the city's airports. The much denser white lines are his journeys on the ground: walking, driving, getting the bus. The big white orbiting circle is Wood driving round London's ring road, the M25, the London Orbital. There is a dense clot of white activity in the heart of the city, a tangle of umpteen walks.

Wood compares the map to a fingerprint because it is so particular and unique. Yet it's also a map of commonplace routines. It traces the precise travels of one man but it captures the back and forth of all our lives. We all follow, pretty much, the same routes. Mobility and lifestyle are made vivid, lasered in yellow and white, yet these bright lines of movement are shared by millions.

Wood explains that he called the map *My Ghost* because 'it refers to a past me. It's not me right now'. It sounds like an obvious point but putting past movements on a map throws up intriguing questions about our relationship with maps. Maps show possibilities: where we could go and what we could do. They are empty. They don't show the filled, busy use of all those streets. I think Wood is telling us that we inhabit not just the map but the city like ghosts; our insubstantial motions temporarily haunt the fixed stuff, the roads and buildings, the brick and concrete, but just for a while and soon our steps and noise are gone and forgotten.

It's a simple map of simple lines but it has a profound appeal. The way it captures both the personal and the impersonal – of our presence in the city but also our ghostliness – means that it captures an anxiety that threads its way through modern life. Don't we want to be seen; to assert our special identity, to express ourselves, to declare our individuality? The idea that these things are important is drummed into us at every turn. But we know and are everywhere reminded that we are one in many millions all doing the same, taking the same routes, probably thinking the same things. Our precious 'identity', our 'fingerprint' on this world, is real but transient and weightless.

Wood loves maps because, he tells us, they 'say things that words and pictures can't'. He quotes the quirky German historian and travel writer W.G. Sebald as an inspiration. Sebald's travel books take us to ordinary places, such as near where he lived in eastern England, and defamiliarizes them and makes them strange. For Wood one particular quote from Sebald captures what he himself is trying to do with his art: 'The greater the distance, the clearer the view,' says Sebald, yet from above, he adds, 'all knowledge is enveloped in darkness. What we perceive are no more than isolated lights in the abyss of ignorance, in the shadow-filled edifice of the world'.

The bright white loop of the M25 (the large white circle that, at a distance, encompasses the central white network) set against a black background that we spot in Wood's map reminds me of Sebald's insight but also of another travel writer with a fascination with the peculiarities of ordinary places. *London Orbital* by Iain Sinclair is about his walk around the byways of London's outer ring road. It full of eerie presences: Sinclair is drawn to the overlooked, the hidden and the perverse. Most perverse of all is trying to walk in a landscape that is so hostile to walkers. Calling in at the home of the dystopian novelist J.G. Ballard, Sinclair explains that he wanted to 'pay homage' to Ballard, who he says 'has defined the psychic climate through which we are travelling'. Instead of a blessing Sinclair receives bemusement: 'We do have buses in Shepperton' Ballard tells him. Ballard is at a loss to understand what Sinclair is up to. We might sympathize but, like Wood's hard-won map, there is something compelling about walking in the city's busy but overlooked margins.

For Sinclair walking is a kind of rebellion. His ramble around the M25's noisy edgeland is a journey in and against the contemporary landscape. He calls it a 'a provocation. An escape. Keep moving, I told myself, until you hit tarmac, the outer circle. The point where London loses it, gives up its ghosts.'

Pacing this unloved landscape is also a kind of ceremony. There is a touch of magic to it. Like Wood's map, it's transformatory. What does

it conjure? Humanity? Ghosts? Sinclair calls the huge roads that circle London a 'presence that nobody wants to confront or confirm' and explains his mission as an 'Exorcism, the only game worth the candle'.

Congestion at Junction 2 of the M25

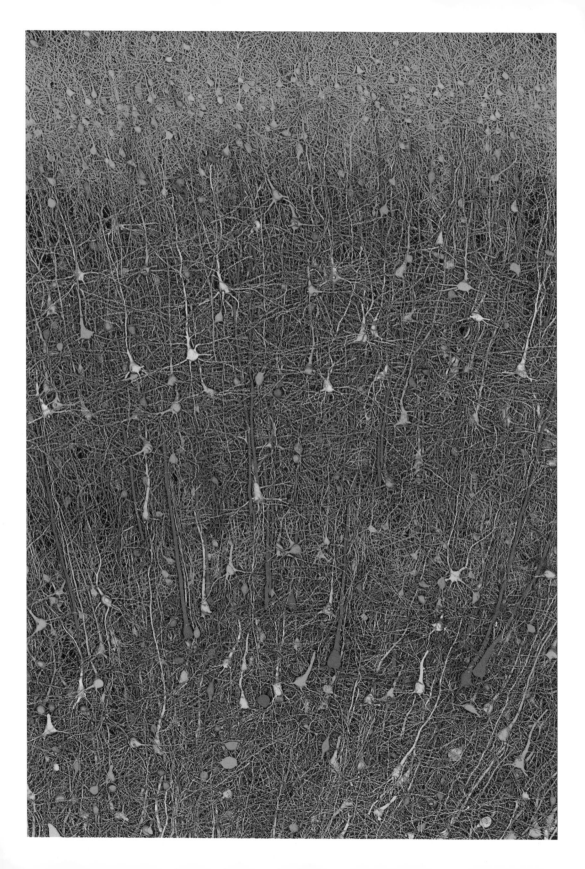

Map 26

Scanning Electron Microscopy Shows Individual Neurons in a Piece of Cerebral Cortex, 2.5 mm tall, 1 mm wide, and 165 microns thick, 2022

Mapping the human brain, one sliver at a time

Every thought, every action comes from the brain but our knowledge of it is limited. We have rough ideas, broad theories, but for a long time it was a black box. That box is now being prised open and light is being shone into its darkest corners.

Previous maps of the brain were general and regional. Arrows pointed to one lobe that does memory; another that does sight. By contrast, the image captured here is at the cutting edge of science. It's not just more detailed than previous iterations, it begins to reveal a total architecture: the brain in all its profound intricacy.

Human brains contain eighty to a hundred billion nerve cells, called neurons, and it is these neurons and the connections between them that we're looking at in this image. The colours signify the type and character of the neurons and their function, such as transmitting and inhibiting messages. The blue cells are large inhibitory neurons and other colours represent different sizes of transmitting, or 'excited', neurons.

It's at the frontiers of science but it's at a tiny scale: it offers a map of just one cubic millimetre of the brain.

The map was created by scientists from Harvard. The same team had already made headlines by mapping a portion of a fruit-fly's brain. The human brain is reassuringly more sophisticated. We can tell this from the amount of computer memory space the creation of the Harvard team's full three-dimensional image requires, namely 1.4 petabytes. I'd never heard of a 'petabyte' so I looked it up: 1,073,741,824 megabytes. It's a number that is best understood by comparison. All the satellite images of Earth taken over three decades by NASA's Landsat program require less computer memory space (1.3 petabytes) than this square millimetre of brain. It has been estimated that to map all the connections manually, counting them under a microscope, would take one million years.

So far so extraordinary. But we'll never get close to appreciating the scientific importance of this map without knowing basic information about the brain. The brain is arranged into regions and different parts are assigned specific functions. If I have a tricky problem to think about I hold my hand to my forehead; it feels right because this is where we find the brain's frontal and temporal lobes. It's where all the real thinking gets done: my problems are mulled over, hopefully solved, in the region under my hand. It's also where my emotions are created and felt, and memory is stored and consciousness – self-awareness – is housed. Vision is at the back of the brain, just above the cerebellum, which controls posture and coordination.

Everything you do has a location, a home address in the brain. The brain is like a city of billions of office workers, all receiving messages, all passing messages along, feeding into automatic and, occasionally, conscious choices that animate a mindless hulk,

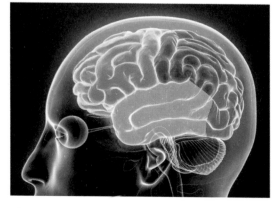

Regions of the brain

the human body. Everything has to be queued up and actioned: when to breathe, when to swallow, when to move, when to worry and when to laugh.

Knowing the regions is essential but doesn't offer detail. Let's try an analogy. Say the only geography we knew of Earth was regional. We have the names and shapes of the continents. Nothing more. How far would that get us? Not far. We would have no idea how to get around Europe, or France, still less Paris. With new maps such as these that all changes: not just France, but Paris is coming into focus, not just Paris but every street and bar; it's like a Google Street View of the brain.

A new era in brain science and medical knowledge is being glimpsed. Dr Alex Shapson-Coe, one of the Harvard team, explains that 'until now, we haven't been able to completely map these connections within even a small region of the brain'. This 'advance opens up the possibility of comparing networks of healthy and diseased brains', Shapson-Coe tells us, allowing us 'to identify the network changes that are thought to cause mental illnesses and other neurological disorders'.

These new maps auger a better understanding and hence treatment of a range of medical problems. One of them is epilepsy. The bit of brain we're looking at is not a generic image. It was donated by a forty-five-year-old woman who suffered from epilepsy and found that no drugs helped her condition. She had to have brain surgery to cure her condition. The malfunctioning bit of the brain, the source of her seizures, was removed and she donated it to the Harvard team. What we are seeing in our map is not 'the human brain' but her brain.

Her donation was stained to make the cells visible under an electron microscope, and then sliced up. It was cut into layers 30 namometres thick (a namometres is one millionth of millimetre). 'You can think of the brain like a bowl of spaghetti', says Professor Jeff Lichtman, who led the Harvard team, 'We sliced it into very thin images and then traced each strand of pasta'.

To chart the inner networks of the human mind opens up the possibility of cures, real hope, but it also poses questions. If consciousness can be witnessed, zipping along in electrical and chemical signals, then the capacity to shape it, at the finest and most detailed level, also comes into view. To know your way round the brain is powerful knowledge.

When new maps that change the world appear it often takes a while before their significance sinks in. This map offers no instant cures, but it is a profound advance. Before this there were generalities; now we have detail. Until very recently, it would have seemed like the most fantastic science fiction to imagine such a thing was even possible. Now it is. We're just at the start of charting the brain in this detail. It is dazzlingly complex. Mapping the whole thing is years away but advances in computational power are transforming the timelines by which discoveries can be made and a full map of the brain is on the horizon.

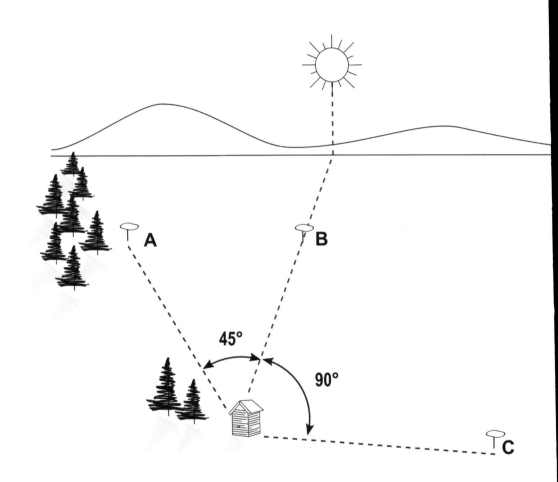

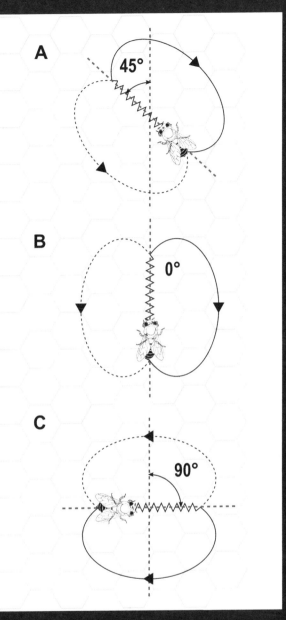

Map 27

Three Waggle Dances Create Three Maps and Show Three Locations, 2017

Dancing geography: how bees draw maps

Let's start with a simple question. How do I get to the park? 'Turn right at the end of this road, and then left at the traffic lights'. Thank you. Nothing has been drawn for me but a map has been made. Hopefully I can remember it and successfully reach my destination.

Let's do that again, but this time I'm a honeybee and you're a honeybee and I want to get to some flowers. No talking this time but you can still map it out. We both know where the Sun is. In the world of bees, as for most flying animals, the Sun dictates direction. To get me to the flowers what you have to do is describe where they are in relation to the Sun. Your map is made by a dance called the waggle dance. Our map translates this bee-geography into human geography. It shows the Sun and three places with flowers, which we call A, B and C. You want to send a bee to the flowers at A? They are 45 degrees left of the Sun so your dance plots a line at a 45-degree angle. The wavy or 'waggle' bit is along that central line and the longer it is the further away the flowers are.

Simple but precise. It's even simpler if the flowers are in the same direction as the Sun. This is what we see for the flowers at point B. No angles this time. The central line of the dance is straight up. In the world of bees this can only mean one thing, Sun-wards. High-speed runs – dashing round quickly to the waggly bit – add even more info: it shows a destination has a lot of nectar or pollen on offer; slower runs indicate that the flowers are less bountiful.

The Sun is associated by humans with blue days and a ball of yellow light. Not so for bees. It doesn't matter how cloudy it is. Bees' eyes are super-sensitive to light and the Sun is a clear guide to them even on the dullest days.

There is more than one type of bee-mapping dance. Another is called the round dance. It's used when the go-to flowers are close by. The bee walks in a circle, back and forth, flipping directions, many times.

These maps are not used by all bees, only by the most gregarious, the honeybee. It has recently been discovered that the bigger the hive the more benefit these locational techniques provide: the larger the colony size and the more dancing there is, the more effective the foraging.

Honeybees are after two things: nectar and pollen. Once a bee finds a good flower loaded with both or either, she returns to the hive, and dances to tell other members of her colony where to find it. The dance is the key but other, more intimate, information is also passed on. Dancing bees will also share a bit of the nectar that has been collected as well as the flower's perfume. These smells and tastes also appear to help other bees know exactly where to go.

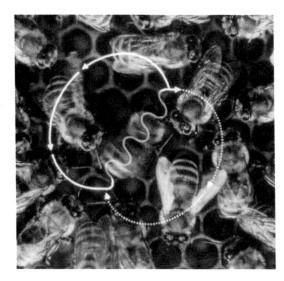

Bee performing a waggle dance

Researchers at the University of London have studied the ways these different bits of information are used by bees. Dr Matthew Hasenjager explains that scientists sought to 'tease apart the effects of following dances from other ways bees can share information about food'. The results provide fresh insight into what we might call 'bee geography'. 'We found that bees searching for new foraging locations relied overwhelmingly on dance-based information', says Dr Hasenjager, 'whereas decisions to revisit known locations' were guided mostly by taste and smell.

Surrounding any dancing bee is an eager audience. Let's not forget that all this dancing and sharing is being done in the pitch dark, inside a hive, and often on the vertical wall of a honeycomb. It seems that wherever they are, bees know the difference between up and down. The surrounding bees don't see the dance. They feel it. Bees' antennae are very acute and feel the smallest of vibrations. It appears that it is, primarily, through their antennae that they experience all the intricate motions of the dance.

It took many years before the language of bees was translated. When Austrian biologist Karl von Frisch first proposed, in the 1920s, that these dances might be how bees found their way to flowers, few believed him. Bees were thought to be simple and haphazard creatures. Sophisticated communicative techniques were thought to be a human preserve. By applying painstaking measurements and using protractors and stopwatches, Frisch was able to overturn this conceit.

It was a revolutionary moment. It opened the question of what intelligence is. It isn't just humans that can describe a landscape and explain its resources. Rather than being an individual attribute, 'intelligence' turns out to be highly social. Following Frisch's work, the notion of 'the hive mind' became popular and, with it, the idea that intelligence arises from the many not the few.

Another small creature that displays this aptitude is the ant. Ants have a very different relationship to the landscape than bees. Individually their movements are, indeed, haphazard but collectively they make sense. Studies have shown how they cope with uneven terrain and soon work out the fastest route to where they want to get to. Ants that have come across something worth getting to – a source of food – lay down a trail of chemicals, which the other ants follow.

Having experienced and rebelled against the horrors of Nazism, Frisch, who had Jewish heritage, had a wry view of human intelligence. A quote often attributed to him is the 'ant is a collectively intelligent and individually stupid animal; man is the opposite'.

The work of Frisch and those who came in his wake changed the way we think about intelligence and communication. But our understanding of how animals navigate and communicate maps remains in its infancy. We know that many animals, large and small, travel with impressive efficiency. In situations where humans would likely be fumbling to get Google Maps to work on their phones, they seem to know exactly where they are and where they need to go.

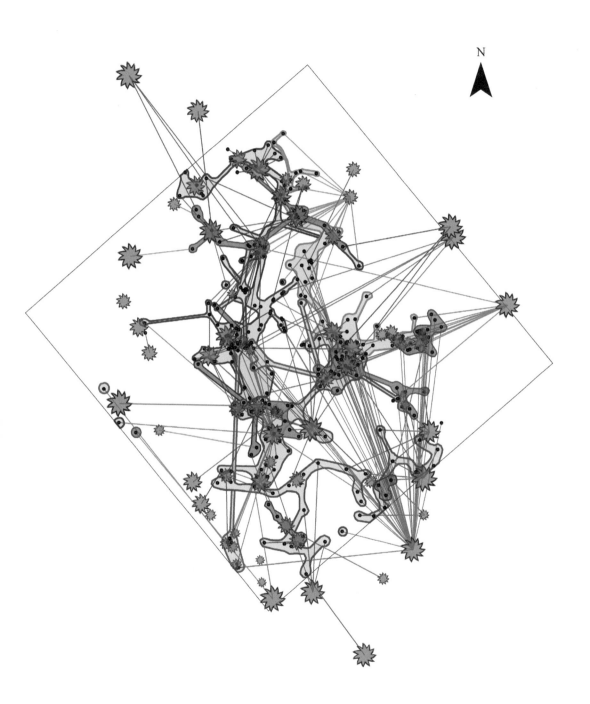

Map 28
A Map of the Connections Between Fungi and Douglas Firs, 2009

Wood wide web: how trees and fungi help each other out

Looking outside of my window, the beautiful trees are in full leaf. They appear to simply loll quietly in the breeze: they are peaceful and utterly mute. Or perhaps I just don't know how to hear them. This is a map of trees with the mute button off.

What we see are lines of communication and connection, channels that send provisions and despatch warnings. It's a simple map but it expands and changes the way we look at plants.

It is a map of a 30-metre square (36-yard square) plot of 67 Douglas firs, each shown as a green-toothed circle. The size of the circles corresponds to the diameter of each tree. Douglas firs are a type of pine tree, native to western North America. They are big trees: up to 100 metres (328 feet) tall. But the map has other tenants too, as indicated by the pink and light blue blobs. We're not just looking at trees but at the fungal network which connects them.

Fungi are the dark matter of the forest. Largely unseen, they hold everything else together. Mushrooms are the visible bit. Mushrooms are the fruiting body, popping up every so often, the tip of a very large iceberg. Beneath them is an extensive network of tendrils that creates a supportive mesh running through the soil, sustaining and linking. Through this fungal network, trees and other plants exchange water, carbon, nitrogen, phosphorus and, according to the latest research, information.

There are two types of fungi on our map, coloured pink and light blue, each a variety of *Rhizopogon* (which means 'root beard') and they are often found living alongside pine trees. The coloured lines show links between fungi and particular trees. The bigger trees tend to have more connectivity, and one, towards the bottom right, we might think of as a mother tree. It is by far the most meshed, directly linked to forty-seven trees. The black dots show the locations where the scientists took their samples.

We have known for some while now that trees and fungi have a mutually beneficial relationship. Fungi weave their way in and around roots: the fungus gets nutrients and water this way but so does the tree. The relationship is so close that the fungal threads become part of the root system, soaking goodness back up into the tree.

Another well-established fact is that plants can send chemical distress signals into the air. If a plant is being chomped by caterpillars, it can let the neighbourhood know. For example, when attacked jasmine generates jasmonic acid, which sets chemicals loose that act as a signal to other plants to prepare for attack. Jasmonic acid also attracts insects that eat the plant's enemies.

Douglas fir forest

In plant ecology circles these intimate ties across species are well-known. What is new is the geography, the networked nature, of these bonds. It used to be assumed that connections were individual: one tree, one set of fungi, one relationship. Maps like this one show something else. Fungal networks appear to connect not just a few or a dozen but hundreds maybe thousands of trees.

This is a simplified map: there are lots more species of fungi, plants, bugs and worms both above and below ground, that, most likely, take part in this intricate web. Science writer Dave Hansford explains that the 'true complexity of the underground network, if all the potential terminals were considered, would be mind-boggling'.

In a controlled experiment, Douglas fir and ponderosa pine seedlings were separated by mesh barriers dug into the soil. It was a fine mesh, stopping any root contact between the two lots of saplings. Fungi, however, can get through most barriers. The first question to find an answer to was what would happen if one set of trees was attacked. It sounds cruel but the researchers pulled off the needles of the infant Douglas firs. Through the fungal network these little trees let their distant cousins across the mesh barrier know the bad news. Stress alerts were sent and the ponderosa pines started to make protective enzymes.

We may think of the natural world as dog-eat-dog. There is a lot of scrapping in the wild but it seems there is also mutual aid. The stripped Douglas firs really went the extra mile. After sending out warning signals they then started sending nutrients to the other trees. Sacrificing themselves for the well-being of another species.

This map shows a mutual support network. Suzanne Simard, one of the authors of the study that created it, has proposed a range of beneficial connections. Older trees help saplings by sending them carbon, for example. Simard tells us that 'These plants are not really individuals in the sense that Darwin thought they were individuals competing for survival of the fittest'. In fact she explains, they are 'trying to help each other survive'.

But let's not get carried away with the idea that trees and fungi are best pals. After all 'communication' and 'connection' are not always about messaging love hearts and posting treats. Plants can also send hate mail. Studies have shown that some plants send out airborne pathogens to stifle the germination of competitors and the fungal network can be used to the same deadly effect. The innocent-looking marigold is one of the worst. It turns out these sweet-faced flowers have a ruthless streak. Marigolds with fungal friends use their network to send out toxins to sicken competing plants and even kill the nematode worms that these rivals are benefitting from. Marigolds can be mean.

There is a widespread desire to find wisdom and goodness in nature. Whether it's poets wandering among the daffodils or films like *Avatar*, where nature – trees especially – is benign and divine, the way we make sense of nature is shaped by sentiment and a kind of desperate hope.

Trees and plants are not kindly grandmas but nor are they wolves. They work together and they work apart. We have lived with them for years but we are only just beginning to fully understand them.

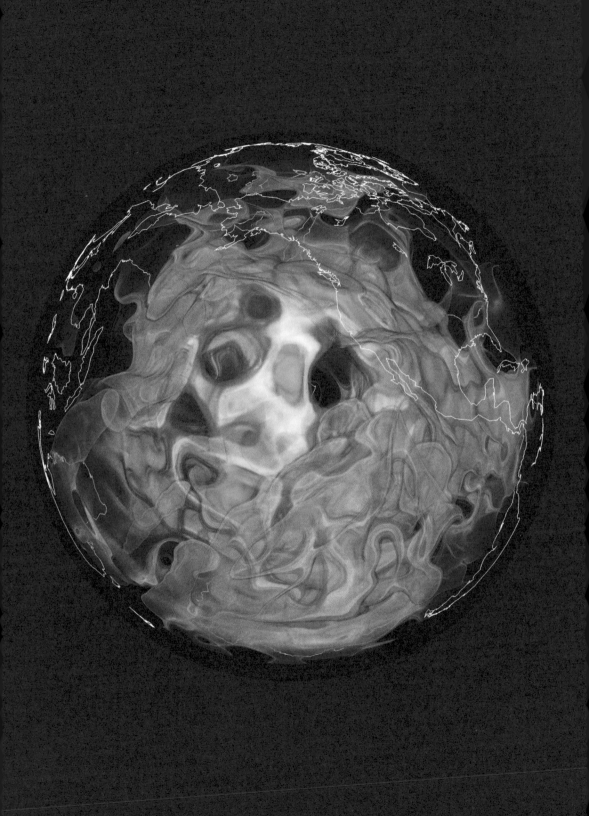

Map 29

Computer Simulations Use the Speed of Seismic Waves from Earthquakes to Reveal the Existence of Subterranean Structures, 2015

Earth power: how earthquakes reveal a hidden planet

These swirls of sun-kissed auburn and butterfly blue tell a terrible story. They map the shorelines of seismic waves, of the huge and catastrophic power of earthquakes. We're looking at the magnitude of seismic speeds: the slower movements are in red and orange and the rapid and dangerous in blue. We are peering into the deepest layers of the Earth to reveal a new vision of our violent planet.

The map looks down on the Pacific ocean. There's a cobalt arc of rapid waves hung upon the edge of the ocean. This energy is clearest along the west coast of the Americas and there's a worryingly bright patch of blue somewhere near Tonga (towards the bottom left of the globe).

This is a crunch point, where the Pacific and Australian plates collide. It is a ferocious zone: in between the big plates a number of geological oddities – microplates – shove and butt up against each other. The Tonga microplate is the world's fastest, moving at speeds of up to 24 centimetres (8¼ inches) per year.

Human bodies, as well as seas and soils have been mapped with energy waves for decades. Sound and radio waves bounce back to give a reading that can reveal the existence of a tumour, the beating heart of a foetus, the location of a submarine or the line of an ancient wall hidden beneath a modern city. The energy of earthquakes can be used in the same way. Our expanding network of earthquake detectors allied with new generation computers pull together data from thousands of tremors.

Under the fragile and comparatively egg-shell thin crust we run around on, lies the Earth's mantle; with a consistency of stiff caramel it's up to 2,900 kilometres (1,800 miles) thick. The map shown here gives a unique glimpse into this vast and, until now, unpredictable subterranean kingdom. It isn't just a map of a seismic speeds, it reveals the beating heart of the Earth.

It's promise and potential are huge. Maps like this can show us the locations of fast-moving magma and so help us know when volcanic eruptions will happen, as well as where and how big they will be. Seismic waves slow down when travelling through underground features, like aquifers, magma chambers and hydrocarbon reserves. This means that water and mineral resources and many other hidden features can also be decoded from these swirling patterns.

The map is the creation of a Princeton professor, Jeroen Tromp and his colleagues. A key member of their team is Titan, one of the world most powerful computers. It's a machine able to work out more than twenty quadrillion calculations per second. What we are seeing in this stunning image is not just the guts of our planet but the centrality of computing to deep mapping. Tromp explains that 'seismology is changing

Tsunami Rock, Tonga

at a fundamental level due to advances in computing power'. We're on a journey towards full visibility, the glass Earth.

Around the world there are numerous small earthquakes every day and every year there are about sixteen huge ones. The damage they do is increasing: thousands more deaths, bigger piles of ruins. These awful sights might trick us into thinking that earthquakes are getting worse or more regular but neither appears to be true. What has changed is us. There are billions more of us than there used to be and a lot of us live in areas, such as Turkey or along the west coast of the USA, where the guarantee of a major earthquake coming along is an iron-clad certainty.

There have been advances in proofing buildings against tremors, but they have had limited roll-out and, in densely packed cities, they can only do so much. In Japan, despite having the highest levels of preparedness and towering defensive walls round the coast, damaging quakes and tsunamis remain not just likely but inevitable.

The most effective measure against earthquakes is what this map points towards, accurate hazard assessment. It's patterned like abstract art, but the real beauty lies in what it promises: fair warning of danger.

We live on a jittery, juddering rock and, like specks of dust on a bouncing boulder, we have no control of its movements. It's a sobering reality. It might make us feel unimportant. Yet my eye keeps coming back to that dense patterning of blues around Tonga because it sparks that most human and fragile of things, a memory. I was on my way to a new volcanic island, whose name was never settled on. It was thrown up out of the sea in this very spot in 2014. I was due to sail there with Branko, a fifty-eight-year-old Croatian-Italian-Swedish-Tongan mariner and local legend. . 'We were the first to come there,' he tells me. 'Ah, that lake! Where ground zero was!' Branko makes a cannon noise and continues: 'It's a green colour; it smells a bit green like green paint, and the ocean is deep blue so you can stand' – now he leaps up, filling my small rented room – 'Yes! With the deep-blue ocean on this side and the green, green lake on the other.'

I'm a wide-eyed child as the island is conjured before me. 'Plants? Yes, yes, plants is already growing. We planted six coconuts and I saw them last week and they are growing. But normal trees, grass have started growing. There is thousands of birds, eggs, little chicks. Everywhere, everywhere on the ground. You can see it from the boat, it's green.'

I never got there. A storm came and Branko's launch was just too small for the grey cliffs of water that massed around us. We turned back. I spent the next few days in Tonga. Each day neatly dressed schoolchildren would run in laughing packs though the lanes; rehearsing, yet again, their mass evacuation plan, for when the next earthquake, the next tsunami, rolls over the low-lying fields and villages of their ancient homeland. Before its name could be settled, Branko's new island blew apart. In January 2022, a massive new eruption sundered it in half and it tumbled back into the sea.

It's a little story from a small country but I'm reminded it of by that pulse of blue. This is not just a map about rocks and raw energy. Within those shapes and swirls are human stories, places that come and go, and children running.

Overleaf: Tectonic plates of the Pacific and locations of earthquakes. The small but highly active Tonga Plate is near the middle.

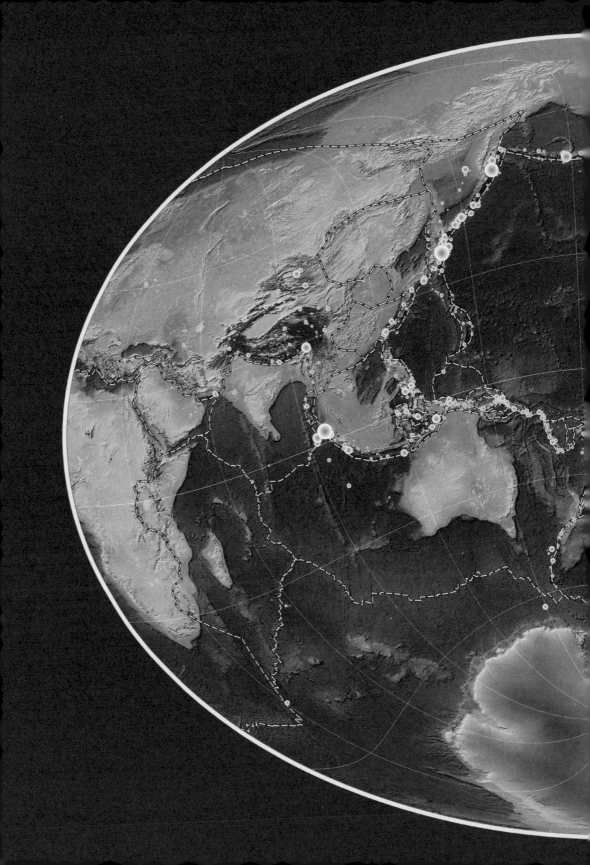

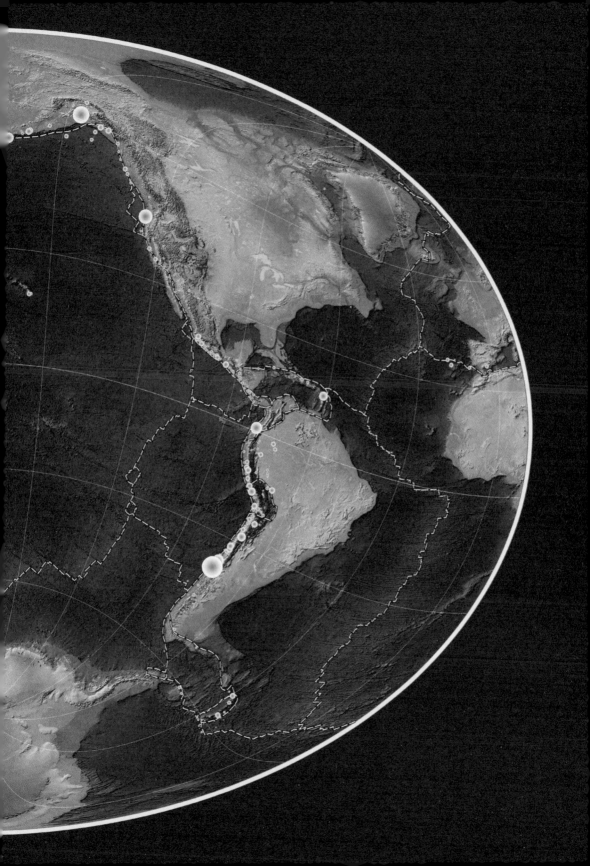

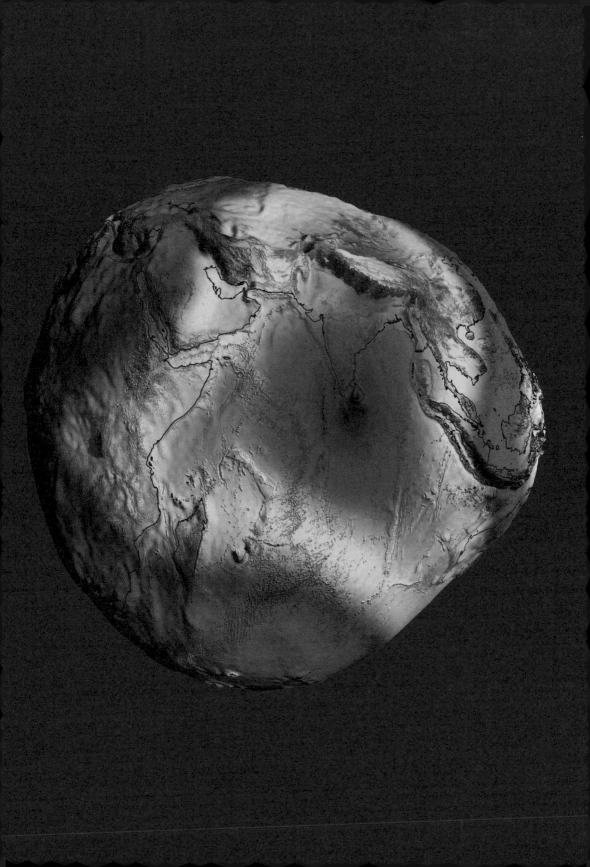

Map 30
New GOCE Geoid, 2011
Planet potato: a map of gravity

The Earth is vegetable-shaped. More potato than sphere. It's fat in the middle, covered in bumps and lumps and a bit squashed. Gravity, the force that makes the apple fall down rather than up or sideways, is determined by mass. For stuff on this planet that means the mass of the Earth. All that planetary unevenness has created different amounts of mass in different places, hence their gravity is different. It all depends upon where you are on the planet.

Our map provides a snapshot of this wonkiness. It's does this by imagining that the surface of the Earth is all water, then showing which bits have more potential pulling power (so that's the higher bits) and which have less. The effects are multiplied, exaggerated massively to demonstrate real differences. The result is what is called a geoid, and may be the most useful map you've never heard of.

Scientists define the geoid in more technical language as the equipotential surface of the Earth's gravity field. It's the shape the ocean surface would take if it was imagined to be under the influence of gravity alone and to stretch across the whole Earth. The colours in the image represent differences in height. The blue shades represent depressions and the reds/yellows represent heights. The geoid rises in mountainous areas because this where the Earth is denser and, hence, exerts greater gravitational pull. On a water-covered world these are the places water would flow to: water would be pulled towards those spots, creating a thicker layer.

This may be as clear to you as a crystal spring or as murky as an outflow pipe. Either way, what matters is that the geoid has multiple real-world uses. It gives us a standard and comparable measure of depth and height. Without it we wouldn't know how high or deep things are. Scientists and surveyors use this surface – which is consistent and known across the planet – to provide precise measurements of all sorts of stuff.

Scientists who want to know what's going on under or on the Earth's surface – for example to work out changing sea levels, the location of faults and the likelihood

The Gravity Field and Steady-state Ocean Circulation Explorer: An ESA Satellite

of earthquakes and volcanic activity – use the geoid. The geoid is also an essential tool in navigation, planning and building.

In 2023, new research using this map investigated its most dramatic feature: that dark blue 'gravity hole' in the Indian Ocean, a place where the sea level dips by over 100 meters (328 feet). It shows that the Indian Ocean is the 'lightest' spot in the planet. There is less mass under that spot than anywhere else. Researchers from the Indian Institute of Science have argued – and it remains a hypothesis – that this 'hole' is the result of streams of hot, and hence less-dense, magma welling up from deep inside the planet, plumes like those that lead to the creation of volcanoes.

Until this global geoid was created in 2011, each country had to go it alone. National geoids were created from multiple different data sources, creating a patchwork of country-wide models. The release of the global map was a major advance in the prediction and measurement of the Earth's physical features.

National geoids are not out of business. They still matter and can offer pinpoint accuracy. Recently the UK's mapping body, Ordnance Survey, used its geoid models to announce that Calf Hill, which is located in Cumbria, was officially not a hill but a mountain. 'When it was last measured in 2010, it measured at 609 metres [1,998 feet]', surveyor Eric Hinds explains, 'which is just short of the 609.6 metres or 2,000 feet of height needed for a mountain'. But 'following the changes to the geoid, the height of Calf Hill now measures at 609.606 [metres/2,000.02 feet]. This minute change has boosted Calf Hill enough to be listed as a mountain'. It's not much of a headline – 'Hill Grows Very Slightly' – but it shows that when you need to know the exact height of things you need the geoid. It also shows you need to keep taking measurements. Neither the Earth's inner structure nor its surface is stable. The geoid keeps shifting and maps will always need to be updated.

The struggle to measure the Earth with precision is a long story. We've known that the Earth is not a simple sphere for a long time. There was something of a war of words between the Royal Society in London and the L'Academie Royale des Sciences in Paris in the eighteenth century over whether it was flat at the poles (London) or shaped like an egg (Paris). Expeditions were sent out across the world to take measurements to prove one theory and disprove the other. It turned out the French egg thesis was wrong and the British flat poles idea was right.

Further complexity was introduced into these models as gravity began to be measured. Ever-more sophisticated instruments have been devised. Which brings us back to our map and to the so-called, 'Formula 1' of satellites, which helped create it; the GOCE, or 'Gravity Field and Steady-state Ocean Circulation Explorer'. Pictured here, it was launched by the European Space Agency, blasting off in 2009 from the Plesetsk Cosmodrome in northern Russia. It carried a highly sensitive gravity device, which could pick up small variations across the Earth's surface to an accuracy of within 1–2 cm (around ¼–¾ inch). The satellite resembles a small spaceship: its thin body and fins allowed it to be highly stable and fly very low. Among its many achievements, it became the first orbiting seismograph, picking up shockwaves from earthquakes on Earth. After running out of propellant, the satellite crashed back to Earth in 2013. Its mission was over but the maps it gave us were game changing.

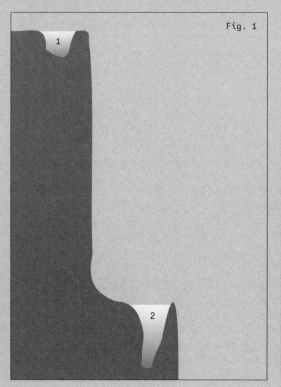

Fig. 1

No.	Name of lake	Elevation (ft)	Depth (ft)
1	Lake Tilicho	16,138	279
2	Lake Titicaca	12,507	922
3	Lake Baikal	1,494	5,387
4	Lake Superior	601	1,333
5	Lake Michigan	577	925
6	Lake Erie	568	210
7	Lake Ontario	243	802
8	Dead Sea	-1,412	997
9	Marianas Trench	0 (sea level)	36,201

Fig. 2

Fig. 1

Fig. 2

9

Map 31
Lakes and Oceans Depth Comparison, 2024

Vertical extremes: from Challenger
Deep to Tilicho Lake

Most maps take a bird's eye view: they peer at the world from above and present a horizontal plain. But in in life we travel up and down not just along. Verticals matter. If you had to build a ladder between the world's darkest depth to its topmost lake, from the Challenger Deep trench in the Pacific to Tilicho Lake high up in Nepal, it would need to be sixteen kilometres (almost ten miles) long.

We're not used to seeing geography side-on. This map shreds the common-sense notion that seas and lakes are basically and broadly at about the same level. They're not.

Our attention is caught first by the chasm on the far right of the map: the one that leads down to the ocean's deepest point. Then we might notice the great height of the lakes shown on the far left. Or looking at the middle plateau, which contains a cluster of famous North American lakes, there is an out-of-kilter depth, Russia's Lake Baikal.

It's time to descend to Challenger Deep, the lowest point of the Mariana Trench. We're going 10,916 meters (35,814 feet), deeper than Mount Everest is high (8,849 metres/29,032 feet). As the water around us darkens and chills we might recall the name of our destination. 'Challenger Deep' pays homage to one of the first attempts to map the oceans. HMS *Challenger* voyaged far and wide between 1872 and 1876 on a scientific mission to explore the ocean's depths. Its work created a new category of knowledge, called oceanography. HMS *Challenger's* discoveries were hailed at the time as revolutionary. In the words of a later pioneering ocean scientist, John Murray, this was 'the greatest advance in the knowledge of our planet since the celebrated discoveries of the fifteenth and sixteenth centuries'.

Stephen Stukins, Earth Sciences curator at London's Natural History Museum, explains that HMS *Challenger* had a 'ridiculous amount of rope on board to measure the deepest parts of the ocean'. It needed it. Wanting to take a sounding in the western Pacific, west of the Mariana Islands, the crew threw over their weighted rope and it just kept spooling out: whipping away in coils and folds and only coming to rest at a stupendous 8,184 metres (26,850 feet). The fact that our planet could be as deep as it is high was a revelation.

The spot since named Challenger Deep in honour of that mission is an even lower point of the Mariana Trench and was discovered in 1951. The US Navy's bathyscaphe (a crewed submersible) *Trieste*, with two crew on board, made the first successful descent to Challenger Deep in 1960. More recent explorations have expanded our appreciation of what were long thought to be dead zones. The pressure of all that water pressing down would splat you and me like an unlucky bug. Yet these pitch-black waters are not lifeless: they are teeming with strange creatures. Over the past two decades expeditions have come back with ever more extraordinary footage of new species. Many are outsized versions of creatures common in smaller form higher up, reflecting the curious phenomenon of 'deep-sea gigantism'.

Driving across Lake Baikal in winter

Lifelessness is a characteristic of very high water rather than very low. Tilicho Lake in Nepal is our highest lake, sitting at an altitude of 4,919 meters (16,138 feet) and absolutely nothing lives in it. Like the surrounding peaks of the Himalayas, it is barren of plants or animals and completely frozen during the winter months. The only life you'll see are brightly clad trekkers who have decided to avoid the crowds at Everest and enjoy a less-travelled wilderness.

The next-highest major lake, Lake Titicaca, which straddles the border between Peru and landlocked Bolivia, could scarcely be more different. It's bustling with activity of all sorts, including the manoeuvres of the Bolivian navy. Human ingenuity has long been present here. An indigenous fishing community, the Uros, used to build tiny floating islands made of reeds many miles from the shore to be safe from warlike neighbours. Each of these artificial islets lasted about thirty years and so they needed to be continuously repaired and renewed. The Uros still maintain their island-building traditions though, today their reed rafts are closer to the shore and many sell trinkets to tourists.

Biodiversity and bustle is not associated with the Dead Sea, which lies well below sea level, a fact starkly revealed on our map. This depth ensures a level of saltiness that kills off most life. One of the few things that can tolerate existence in the Dead Sea is green slime.

Finally, back to the question posed earlier: what is going on with Lake Baikal (lake no. 3)? It's in the south of Russia and, once you know where to look, it pops out at you on any world map, a great blue gash near the Mongolian border. It's got far more water in it than any of the Great Lakes because it's so deep, the bottom being 1,187 metres (3,894 feet) below sea level. It's so deep because it sits on top of a continental rift zone. This means the Earth's plates are coming apart below it and they continue to do so. Baikal is getting deeper with each passing year. This has been going on for millions of years, long enough for numerous species to evolve and adapt to its unusual environment. It is called the 'Galapagos of Russia', and among Lake Baikal's many unique species are giant spiny crustaceans, ghostly white fish and the world's only freshwater seal. Like a lot of deep lakes Baikal is home to myths and legends. The fearsome 'Lusud-Khan' or 'Water Dragon Master' is said to inhabit the lake and to resemble a giant sturgeon, with a long snout and armoured plating on its back. It's a bit like the Loch Ness Monster, though since the surface area of Lake Baikal is about 580 times that of Loch Ness, there are a lot more places for Lusud-Khan to hide.

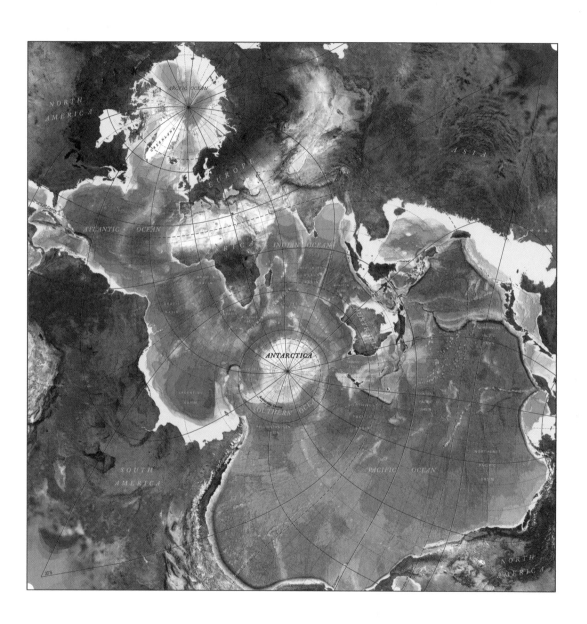

Map 32

General Bathymetric Chart of the Oceans (GEBCO) World Ocean Bathymetry, 2022

There is only one ocean

The world is two-thirds water. It's one of those facts that is easy to repeat but hard to grasp. That's why this map is a revelation. It does a simple thing, it recentres our point of view, and by doing so it turns our world upside down.

Now we can finally see not only how watery the Earth is but how all these seas and oceans are connected. We like to pretend they are separate. We divide them up, give them labels and run borders through them. The truth is in front of us: there is only one ocean. The Earth's countries are arranged on an edgeland that breaks the surface of a globe of water. In fact, all the continents could fit inside just part of this single ocean, the bit we call the Pacific.

This map enhances our sense of water's realm by following the convention that turns Antarctica and Greenland white. These regions are dry land (though see Map 16, page 74) but covered with (frozen) water.

The light and dark blue shadings in the ocean show depth and terrain and point to the creators of this map, an international team of bathymetry experts. 'Bathymetry' combines the Greek words for deep ('bathus') and measurement and is that branch of map making dedicated to underwater terrain. This book is testimony to the many branches of mapping and bathymetry is one of its many exciting horizons. This map, the General Bathymetric Chart of the Oceans, was created in 2022 by the scientists at the leading edge of bathymetry. They call themselves the GEBCO community (see below) and they operate under the umbrella of the UN's Intergovernmental Oceanographic Commission and the International Hydrographic Organization. Less than twenty per cent of the world's ocean floor has been mapped in any detail. There is a lot left to do but the rate of progress is startling. The GEBCO have set themselves the goal of creating a definitive map of the world ocean floor by 2030. This goal is called Seabed 2030 and it's not just about generating new maps. A lot of scattered bathymetric charts exist, generated in many locations and Seabed 2030 is identifying all the parts of this giant jigsaw and connecting them. The idea is that, by 2030, all the world's bathymetric data will be compiled into a freely available global digital map.

This is happening at the same time that satellites are speeding up bathymetry. Traditionally, the seabed is mapped by ship-operated sonar. It's tried and tested but it only maps the thin strip that is sailed. It has been calculated that a full survey of the oceans by ships would take more than 200 years. By comparison, satellites can scan huge swathes. They detect gravity anomalies (based on the geoid; another example of why Map 30, see page 130, matters) and use this data to calculate the topography – the ups and downs – of the ocean floor.

The hope that, one day, the ocean will be mapped has been kept alive by the GEBCO community for more than a hundred years. The group was born in 1903 under the direction of Prince Albert I of Monaco and, from the first, it called forth the kind of international co-operation that characterizes the best science and is required if we are to tackle epic tasks such as this. The founders had an ambitious idea: that bathymetric data collected by ships, regardless of their national origin, would be brought together in one series of maps made available to everyone. It was to be a collective endeavour with collective benefits. And it worked: the data started to flood in from all corners of the world. Various printed and now digital editions of the world chart have been produced, each one revealing new landscapes, unseen canyons, plateaus and mountains.

Most of these features are nameless and the GEBCO has also become the compiler, registrar and arbiter of undersea names. If you want to submit a name for a newfound bit of the sea floor you can write to them and they'll consider it. Off the Atlantic side of the British Isles are a range of features long since named: Porcupine Bank, Porcupine Seabight, Porcupine Abyssal Plain, Hatton Bank, Rockall Plateau, Feni Ridge and Hebrides Terrace Seamount. A seamount is an underwater mountain. Most of the world's volcanoes are seamounts. Our inset map, pictured here, shows Zealandia, which we can think of as a vast upland region with numerous submerged features. There's the Chatham Rise, that arc of highlands stretching out to the right of the South Island, towards the Chatham Islands. Other features include Lord Howe Rise, Norfolk Ridge, Challenger Plateau, and across and beyond these features about seven hundred unnamed, individual seamounts.

The oceans are coming into the light, their secrets and their importance are being revealed. Given its size, it is no surprise that the ocean is the key player in the Earth's climate. It creates the weather and it also stores huge quantities of carbon dioxide: over twenty-eight times more than the quantity of carbon stored by land vegetation and the atmosphere.

It is in the sea not on the land where you find most animals: seventy-eight per cent of animal biomass lives in the marine environment (and ninety-one per cent of species in the ocean still await description). However, because of the size of trees, if we measure the *weight* of all living things, most of it can be found on land (eighty-six per cent of the planet's total biomass). To put it another way, the land is the kingdom of plants, and the sea is the kingdom of animals.

Zealandia: a largely submerged microcontinent

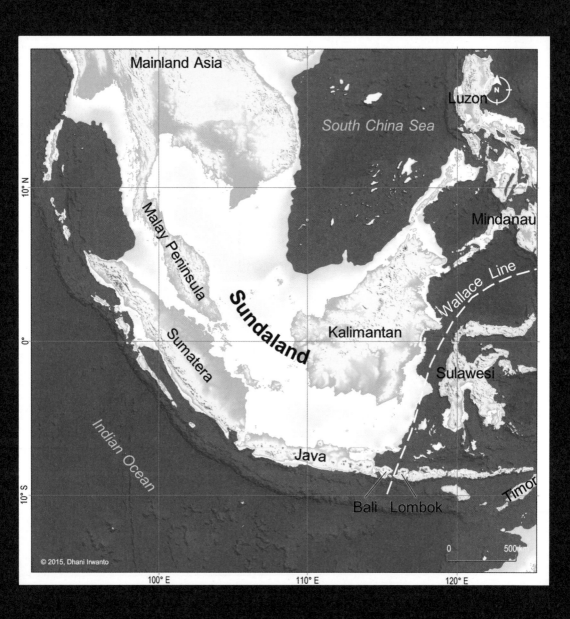

Map 33
Sundaland 20,000 Years Ago at the Last Glacial Maximum Period, Dhani Irwanto, 2015

Sundaland: a hidden ancient homeland, now under the sea

This is one of the cradles of humanity. Generations lived in Sundaland. It is a point of origin: much evidence of early hominids comes from here. Today it's a different sort of hub. The seas covering it are the world's busiest shipping lanes: two-thirds of China's exports travel over this forgotten kingdom.

The map is a snapshot from twenty thousand years ago. It was much colder back then. Ice sheets to the north and south had sucked up a huge amount of water, dramatically lowering sea levels. The shoreline was 116 meters (381 feet) below where it is today. Countries that are now separated by water were connected. Over many generations, before and after the time of the landscape shown in this map, people could walk from Asia over to Indonesia and onto the Philippines. And some of them then crossed the short channel to Australia. It wasn't far but, even then, this was deep water. Humans managed to cross it and Australia was peopled. But it was an uncrossable boundary for other animals. This boundary is called the Wallace line (marked by the dotted white line on the right hand side of our map), and it's an evolutionary watershed: to either side we find species evolving separately. Marsupials abound to the east of the line but there are none to the west.

Humans first arrived in Sundaland about seventy thousand years ago. They were not the first 'people' to make it their home. Other traditions and cultures preceded them. Java Man, a type of *Homo erectus*, had already been living on the savannahs that covered the region for 1.4 million years. The island of Java was then a highland region rising up from a vast expanse of plains and prairies and it's in caves on Java that geometric engravings on shells, scratched with shark teeth by non-human hands, have been found and dated to 500,000 years ago. Java Man was not alone. There were also the Denisovans, creative hominids who, recent finds suggest, may have fashioned jewellery to adorn themselves. Even more unique is *Homo floresiensis*, the so-called 'hobbits', a tiny cousin of modern humans at just 1 meter (a little more than 3 feet) tall. All gone, all vanished, except for bits of tooth, broken skulls, drawings on shells, hints and puzzles. Whether these ancient tribes had any contact with our ancestors remains a mystery.

'Mystery' is a good word for Sundaland. There must be so many things down there, deep at the bottom of the sea: dwellings abandoned tens of thousands of years ago; carvings, jewellery, stone tools, scattered remnants of other lives. To this day, this underwater kingdom remains virtually unexplored. All the information we do have is deduced from discoveries on the islands that formed once the ice sheets melted and the sea level rose. We can get a hint of what might be waiting for us by looking at Sundaland's distant European cousin, which is called Doggerland and is now covered by the North Sea. Even though it's archaeologically less important, it has begun to be explored. Fire pits, extinct animals, finely worked spears have all been photographed, measured, labelled. And something even more telling: shamanic sites thought to be designed to ward off climate change.

Java Man (Homo erectus erectus)

In Doggerland, as elsewhere, the turning of the tide, the rise in sea level, was not slow and gradual but sudden and calamitous. A tsunami swept everything away. Professor Steven Mithen explains how 'many lives were lost; those hauling up nets from canoes, those collecting seaweed and limpets, children playing on the beach, the babies sleeping within bark-wood cots.' Flints and antler hoards found in the construction of Europoort harbour in Rotterdam may have been placed to propitiate the unfolding disaster: treasures offered to and on the shore, to placate the angry spirits of the sea; magic versus climate chaos.

In the absence of facts about Sundaland, myths and speculations abound. It must have been the site of similar dramas: peoples, of diverse species, would have witnessed the loss of their ancestral communities. And they would have tried to get on the right side of whatever force they imagined was causing these changes and adapt themselves to living on their islands.

Underwater archaeology is taking its first steps in Europe but in many of the countries surrounding Sundaland it is waiting to be born. In the absence of facts, myths and speculations abound.

The most plausible of these speculations is the Out of Sundaland theory. It tells us that it is was from here that the diverse populations of Asia and Australasia evolved and spread. The idea of Sundaland as one of the cradles of humanity is both appealing and credible and will tempt future generations of marine archaeologists.

However, currently Sundaland's main claim to fame relies on more fantastic ideas. It is claimed to be the site of the lost civilization of Atlantis. A website dedicated to the many versions of this idea currently in circulation explains that:

> The Sunda Sub-Oceanic Plain is large enough to match Plato's description of Atlantis. Its topography, climate, flora and fauna together with aspects of local mythologies, all permit a convincing case to be made to support this idea.

The Atlantis theory has been doing the rounds for a long time. Thomas Stamford Raffles, the founder of Singapore, first offered it up in 1817 and it has been recycled, largely by amateur Indonesian historians, many times since, as illustrated by Danny Hilman Natawidjaja's 'Plato Never Lied, Atlantis is in Indonesia' and Dhani Irwanto's 'Atlantis: The Lost City is in Java Sea', and numerous similar claims. Other authors tell us Sundaland was the site of the Garden of Eden or once covered in pyramids, whose pointy tops will one day be discovered poking through the seabed.

These are eccentric tales but they give voice to a deep and powerful desire to rebalance the world's origin stories. Indonesia and south-east Asia more generally have been overlooked in global history. In my school history lessons they were not mentioned, not once. We can be fairly sure that the drowned sands of Sundaland are not littered with pyramids. Yet, if we are interested in the peoples and peopling of the Earth, this is not a place that should be passed over. Humans and their close cousins made this their home over immense spans of time. Today their watery tombs are silent witness to changes in climate and sea level that overwhelmed their world and threaten to overwhelm ours.

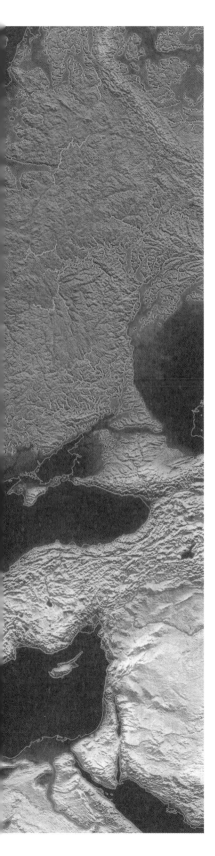

Map 34

Speculative Map Depicting Europe After 80-metre Rise in Sea Level, Christopher Bretz, 2018

Europe in a world without ice

The coasts of Europe have disappeared. In this map the sea is 80 metres (262 feet) higher than it is today. Denmark, the Netherlands, Belgium, the north of Germany and Poland and all the Baltic states have gone. Italy is whittled down to a central spine and Britain and Ireland are archipelagos. The mountainous Spain and Turkey, even after this sea rise, remain recognizable.

What has gone are not just low coastal areas but the best farming land. Southern Ukraine, northern France, the south and east of England as well as Italy's 'breadbasket', the Po Valley: all under salt water. At the bottom of the map, where the Nile is usually seen feeding into a fan of rich farmland, there's little left but desert. This is a map of lost places but also of the rocky and barren places that are left.

It was created by artist and illustrator Christopher Bretz and it's a powerful vision of what Europe would look like if we let rip and burned all the fossil fuel we could find. This rise is a theoretical limit, a figure arrived by calculating what would happen if all the world's ice were to melt. That kind of rise would mean the clock face of Big Ben would be underwater.

I hesitated before including Bretz's wonderful map. If we use up every last bit of fossil fuel and wait for hundreds, maybe thousands of years, it's credible. An understandable reaction on hearing this timescale is a sigh of relief. Big Ben is safe and so are we. This is false comfort. What is coming our way in the next few decades is likely to be about 0.3 metres (one foot) of sea level rise. It's what we can expect for 2050. We're on course for more than double that by 2100.

These rises sound small but they will have profound consequences. When you're dealing with something as big as the ocean, a rise of a third of a metre is a whole lot of water. Rises of this scale present a threat to coastal cities – London, Lagos, Bangkok, Mumbai, Shanghai, and New York – and delta countries like Bangladesh. To understand why we need to know that the problem of rising sea levels is not like a slowly filling bath: steady, calm and predictable. A warming world means surges: ever fiercer storms barrel ever more water inland. With more water and with bigger storms, such surges become difficult to manage. Indonesia has begun to move its capital from the island of Java to safer ground on Borneo, while millions of coastal Bangladeshis, fed up with losing all they have every rainy season, have upped sticks to swell the 23 million living in the city of Dhaka. These are not predictions, they are underway. For my book *The Age of Islands* I visited numerous islands around the world threatened by rising seas and sailed over umpteen lost ones that were dry land ten years ago. I also went to islands where farming has almost collapsed as salt water has been pushed inland.

Eastern Scheldt Storm Surge Barrier

This map represents Europe in a world without ice, but it what it shows of North Africa is just as significant. The loss of all the rich farmland surrounding the mouth of the Nile exaggerates losses that are already happening. This area is currently experiencing rises of millimetres per year. It doesn't sound much but it has a big impact. Costal soils are being poisoned and plants are faltering: thirty to forty per cent of the soils of the Nile delta are now classified as salt affected. This means lower yields and less food going to market. It means Egypt needs to import more food, it needs to spend more money on food, and it's rising population feels the pinch, becomes restless. Even millimetres of sea level change can matter.

About 680 million people live in low–lying coastal zones. They won't all be under water any time soon, but current rates of ice melt suggest that evacuation plans need to be in place. In early 2019 the ominous news came that the Greenland ice sheet is melting four times faster than previously thought. 'This is going to cause additional sea-level rise,' says the report's lead author, Michael Bevis, a professor of geodynamics at Ohio State University, adding, 'We are watching the ice sheet hit a tipping point.' Bevis is sombre: 'The only thing we can do is adapt and mitigate further warming – it's too late for there to be no effect.'

Eighty metres is the ballpark figure for how much the sea would rise if all the world ice were to melt. It's not an easy calculation because lots of factors are in play. One is that in a warmer world the sea expands. Water expands by about four per cent when heated from room temperature to its boiling point. The oceans are not going to boil but, given that two-thirds of Earth's surface is water, even a few degrees of heating is going to add a lot of volume.

Another component is the rise and fall of land. Around twenty thousand years ago ice sheets covered northern Europe and North America. The weight of that ice pushed the Earth's crust down by around 500 metres (1,640 feet). Now that most of this ice is gone, the Earth is readjusting itself; it is bouncing back. This is why new islands are appearing in the Gulf of Bothnia, the scoop of water that separates Finland and Sweden. The bad news for land further south, which is where most Europeans live, is that as the north goes up it goes down, see-saw style. The southern half of Britain is titling downwards, an 'isostatic adjustment' that adds between ten and thirty three per cent on existing sea–level rise.

What can we do? Plenty. Some countries are doing more than others. The Netherlands has centuries of experience of fighting the sea and it's here you find huge defensive structures like the Eastern Scheldt Storm Surge Barrier (pictured here). It's a 9-kilometre (5½–mile) long sea dam that connects delta islands. The dam connects islands but also creates them. On one of these two artificial islands is a slab announcing that: 'Here the tide is ruled by the moon, the wind and us'. It's a mighty boast but we haven't been very wise rulers and, big and long as it is, this dam may not be big or long enough.

N

70 km

Bilbao

Vigo

Madrid

Lisbon

Córdoba

Granada

Seville

Málaga

Cádiz

Gibraltar

An

Valencia

Oran

Map 35
Land Surface
Temperature, 2003

Cooking Iberia: a map of extreme heat

My recipe book tells me to cook my 'steak to no more than 60°C (140°F) in order to retain the most tenderness, flavour and juiciness'. Here's a map of Spain and Portugal that shows the ground is hot enough to cook steak. The black colours radiating off the southern regions are where the land is at or hovering around this mark. Apart from the strip of coolness along the northern coast, the soil temperature is searingly high everywhere.

This is not a map of air temperature but Land Surface Temperature. It's what you would feel if you touched the soil. Which can be summed up in one word: 'ouch!'. Soil is usually warmer than the air but it's no surprise that the air temperature was also setting new records. Air temperatures in the black zones were in the forties (more than 100 degrees Fahrenheit).

This image is from July 2023 and was taken by the Copernicus Sentinel-3 network, the Earth observation satellites of the European Space Agency. They provide vivid evidence that summer heat is now extreme. Most of the Iberian Peninsula – Spain and Portugal – has always been hot in summer but nothing like this.

Why does Land Surface Temperature matter? Surface animals survive by eating what grows in the ground. When soil is, literally, cooked it dies: seed germination stops and the worms and micro-organisms that enrich it cannot survive. The soil falls apart and crumbles into dust, which then blows away. In the coming centuries, one of the top priorities for the survival of *Homo sapiens* will be looking after the planet's soil. When it comes to survival, gold is valueless and dirt is treasure. Only about three per cent of the planet's surface is suitable for arable production and it takes years – some experts say as much as three hundred years – of sifting and decomposition to create just one centimetre (a third of an inch) of fertile soil.

The heating of the world is not an even process. Spain and Portugal, along with other Mediterranean countries, are heating up faster than the global average. Summer 2023 saw a heatwave across much of Europe, making living conditions intolerable. Valencia, a port city on Spain's east coast, recorded its hottest day, nearly 47°C (117°F). 'This summer has been horrible' complained one resident to journalists; 'we cannot live in our apartment'. Others chimed in: 'at night we use the air-conditioning and a fan', 'during the day we try to leave the house as late as we can'.

Firefighters work to contain a wildfire in Pumarejo de Tera near Zamora in northern Spain

Air-conditioning used to be a luxury in southern Europe but now it feels like an essential. The municipal government in Valencia opened an air-conditioned shelter for the homeless. It was lifesaver but air-conditioning is not cost free. It requires energy to manufacture and run and conditioning units cool rooms by pumping heat out elsewhere, mostly onto the streets, raising city temperatures even higher.

One of the immediate threats that extreme heat brings is wildfire. Today the fire season in Portugal and Spain begins as early as March. It's in Earth's 'temperate zones', such as Europe, that wildfire is the biggest risk. This may seem odd. Europe is not one of the planet's hottest spots. You can find much higher temperatures in the sands of Arabia and the Sahara. But there are no trees so there is nothing to burn. You may die of thirst but the fire risk is small. Cooler places are lush with forest, places like Portugal and Spain. They're packed with flammable material. It's a painful paradox: trees produce oxygen and have a key role in reducing pollution, and chilling the climate, at every scale. They are vital to our existence but today any city surrounded by woodlands is a city potentially circled by fire.

Each year is different, but the trend is clear. 2022 was a year of huge forest loss in Spain: 306,000 hectares (756,142 acres) were burnt. 2023 was less disastrous but, across Europe, saw about 470,000 hectares (more than a million acres) of land lost to fire, with a financial cost of more than four billion Euros. Greece was the worst affected, with Spain second. Forest fires were also intense in Turkey, Algeria and Tunisia.

All these countries are scrambling to learn how to cope. In Spain new strategies are being adopted. Rather than throwing water on every fire, firefighters now pick and choose. They allow fires to burn in harder-to-reach locations. Sacrifices are made. They know that it is better to keep water and labour power in reserve, so villages, towns and cities can be saved. With temperatures this high, resources have to be husbanded.

All sorts of fire safety policies and codes, are rolling our way. Replacing pines and other softwood trees with slower burning hardwoods and creating more fire breaks will help. In the USA a new industry has developed around telling people how to manage their 'Home Ignition Zones', the concentric rings – 5 feet, 30 feet, 100 feet and 200 feet (1.5, 9, 30 and 60 metres respectively) – designated by state authorities. Each zone comes with a to-do list, such as 'keep your roof clear of leaves, needles, and debris' and, at 30 feet (9 metres), 'remove all scattered trees and keep grass mowed'. The cost for these is small but there are plenty of bigger bills and the biggest of all is insurance. The market for fire insurance has stopped functioning in large swathes of the USA. Millions of people now live in places too risky to cover.

People in Spain and Portugal have been suffering from intense summer heat but this is not a map about humans. It's about the land. Spare a moment to think about the life in our soil. At 60°C (140°F). We call it 'dirt' but we disregard it at our peril.

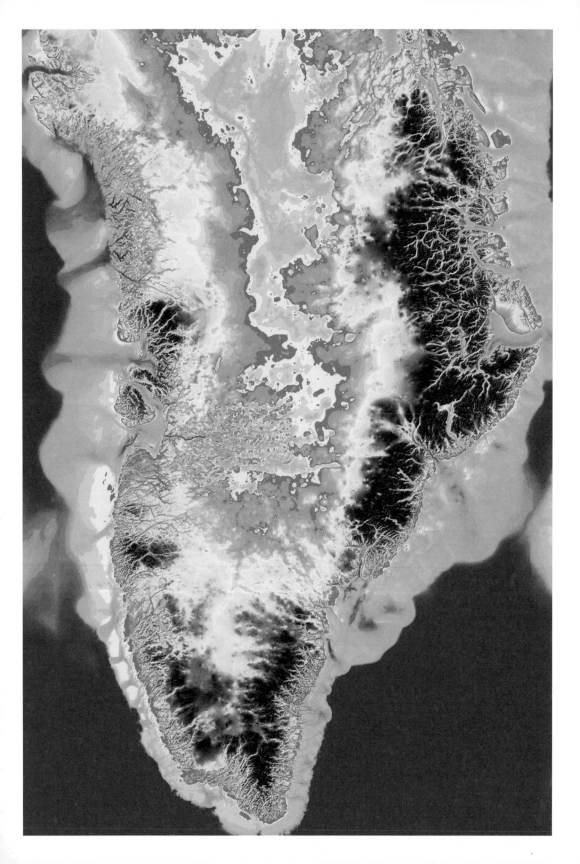

Map 36

Complete Bed Topography and Ocean Bathymetry Mapping of Greenland, M. Morlighem et al, 2017

The hidden great lake

The world's biggest island is covered with a slab of ice, many kilometres deep. It is only at its wave-lapped edges that we get to see Greenland's stunning mountains and plains. But what would Greenland look like if all the ice disappeared?

For all of human history, no one knew. Would it be a scattered archipelago, or one big island? Thanks to this map we have the answer. Greenland is revealed as a single island but an odd looking one: it's a clunky necklace looped around a tremendous lake.

It's not a surprise that the east and south are shown to have ranges of high mountains and steep fjords resembling the coast of Norway. But there are also surprises here. The island's inland lake would be a unique place, freckled with thousands of small islands and unparalleled in size. It would be the world's biggest freshwater lake. You could put the whole of the UK in there and still have plenty of elbow room.

This map is the work of many generations. It was drawn by compiling radar soundings made since the 1960s. An international team, led by researchers at the University of California, created it. Why did they bother? The answer is climate change. Because of rapid ice melt, the race is on to understand the hidden landscapes of the poles. This map gives us a heads up. It tells us, for example, which regions of Greenland will lose their ice first and how and where rising seas will invade inland. The relatively warm water of the Atlantic Ocean won't flood the great mountains over to the east and south, but it is likely to make deep inroads elsewhere. On the western lowlands ice and sea will meet, accelerating ice-melt into the oceans.

The scientists behind this map explain that 'west Greenland and major glaciers in northern Greenland' risk being melted by Atlantic waters, and that these glaciers 'will remain exposed' to invading sea water 'for tens to hundreds of kilometres as they retreat inland'.

There are other lessons that have been gleaned from this map. It shows that many more coastal glaciers are at risk of rapid melting than previously thought. Since it shows altitudes and, hence how thick the ice is that still covers Greenland, it also tells us the volume of water that melting will release. Thanks to this map, we now know how much higher the world's oceans would be in the event that all Greenland's ice melted. The answer is 7.42 meters (nearly 24½ feet).

Another way of looking at this map is as a snap-shot of the deep past. Hundreds of thousands of years ago Greenland was empty of ice. We can now see what it used to look like. How do we know it was once free of ice? Because, back in 1966, US scientists drilled through the ice sheet and pulled up a long-frozen core. It was, mostly, just a tube of ice but at its end it held a surprise. At the end of the tube was 3.5 metres (11½ feet) of soil. Recently this core, long neglected, has been re-examined and leaves, moss and even twigs have been found in it: an ancient ecosystem and proof of an ancient ice-free Greenland.

The story of that 1966 US mission is a curious one. Back then, as now, polar science was mixed up with politics. The US's government's Project Iceworm transported soldiers, and heavy excavation equipment, even a nuclear reactor, to northwest Greenland in order to create a cave system, an ice labyrinth, which they called Camp Century. The idea was that these caves could be used to conceal nuclear weapons from the Soviets. The 'scientific mission' with its ice-cores and heaps of data was a cover for Cold War manoeuvrings.

Greenland is a landscape of secrets. There is so much still to find out about it, but it has a rival in mystery. Antarctica's topography is also nearly all buried under ice. It too has visible mountain ranges near the coast, and these include active volcanoes such as Mount Erebus. New maps, made in the same way as our map of Greenland, are beginning to show what lies beneath Antarctica (see below).

Antarctica has a complex geography, which will shape, accelerate or impede the melting of its ice. Newly discovered high ground will, it is thought, prevent the rapid retreat of melting glaciers but smooth and sloping plains will hasten it. One of the team members of the group that led this research, Emma Smith, uses this analogy: 'Imagine if you poured a bunch of treacle on to a flat surface and watched how it flowed outwards. Then pour the same treacle onto a surface with a lot of lumps and bumps, different slopes and ridges – the way the treacle would spread out would be very different. And it's exactly the same with the ice on Antarctica'.

The Antarctica map also revealed the deepest point on continental Earth. It's a trough under Denman Glacier; an ice-filled canyon that reaches down 3.5 kilometres (almost 2.2 miles) below sea level.

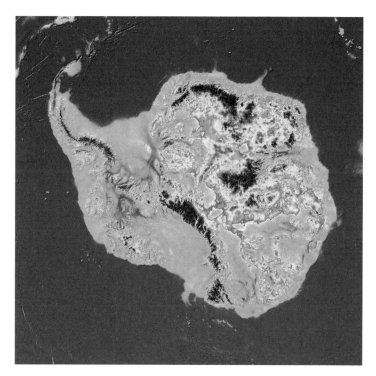

Beneath Antarctica's ice

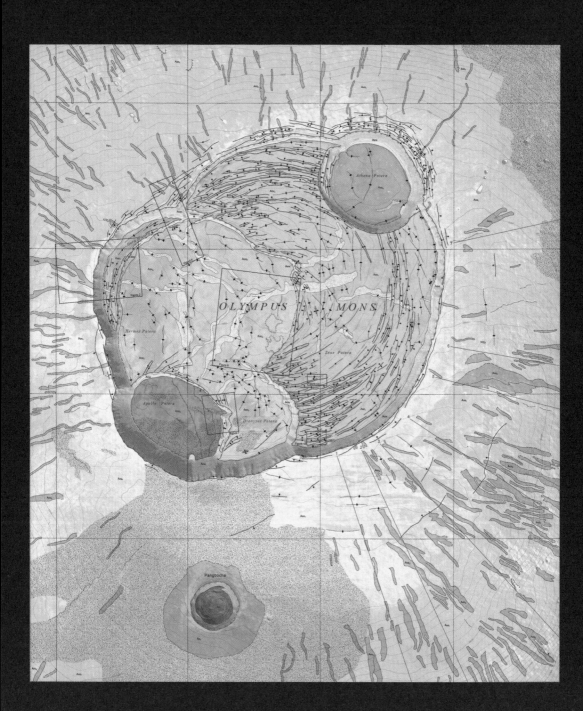

OLYMPUS MONS

Athena Patera

Hermes Patera

Apollo Patera

Dionysus Patera

Zeus Patera

Pangboche

Map 37

Geologic Map of Olympus Mons Caldera, Mars, 2021

Olympus Mons and new maps of Mars

Among the great artists of our era I nominate the graphic staff at the Astrogeology Science Center in Flagstaff, Arizona. Candied lilac kisses peppermint green, swirled round with steep golden bluffs: Mount Olympus, Olympus Mons, is not just being drawn it is being conjured.

These colours are poured by the Flagstaff team onto the wild highlands of the solar system's highest known mountain. Olympus Mons is huge. Its size is only hinted at by the elevations we see here. Around the central caldera those honey-brown slopes and gullies flow down and down. This is no day-trip, no weekend expedition. Olympus Mons is 600 kilometres across and 26 kilometres high (373 and 16 miles, respectively): about the same size as Poland and three times the height of Everest.

This map of Mars is the most detailed we have. It's so fresh, the contours, cliffs and rivulets so complete, that it's hard to believe it's a map of a planet hundreds of millions of kilometres away. It's at 1:200,000 scale; that's the scale we use for regional maps back on Earth like regional maps of France, which use the same scale.

Olympus Mons is a volcano. Twenty-five million years ago lava spewed from its maw. It is thought to be dead now. That makes Mars a safer, less active, planet than Earth. Yet, although volcanic events on Mars are less frequent than they are on Earth, when they do happen they are extremely violent and calamitous. In low gravity, lava spreads rapidly and easily, galloping across the landscape.

Olympus Mons is so big that it's visible from Earth with a simple telescope. It was a key feature of the map of Mars drawn in 1877 by the Italian astronomer Giovanni Schiaparelli. He also sketched the planet's wayward 'channels' (in Italian *canali*). *Canali* was mistranslated into English as 'canals', a miscommunication which fuelled a lot of speculation about Martian engineering.

The map's bright colours highlight different ages and formations. The two smaller round shapes at the top and bottom of the mountain are 'collapse pits' and they speak of ancient drama: how this great mountain collapsed into itself. Olympus Mons's huge eruptions emptied out the magma chambers buried beneath the volcano. Suddenly there were great holes under the mountain.

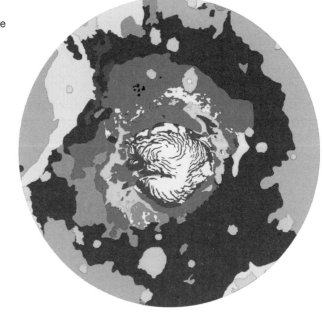

The North Pole of Mars, based on surveys from the Viking Orbiter Spacecrafts

They didn't last: the weight of the volcano took over and immeasurable tonnes of Martian rock fell back down into the void.

If we travel 5,000 kilometres (3,107 miles) west of Olympus Mons we find ourselves in a landscape of canyons and dry ravines. It's called Aeolis Dorsa, or 'Wind Ridge' and it looks like a river system because it is a river system. A clear watery line wends through the valley floor, evidence that water once flowed free and plentifully across the surface of Mars. Mars was once very wet and it had a lot of river systems. The rivers deposited debris downstream, creating fans of deposited material. Across myriad epochs the landscape of Mars was sculpted by water. Water is the basis of life. The idea that simple creatures of some sort lived in and around these streams is a reasonable bet.

Life on Mars is now long dead. All we can expect to find are petrified relics. The rivers dried up eons ago and today, as the name Aeolis Dorsa tells us, it is only wind that whips motion into these dead hills. The wind has been blowing so long and so hard that this river is not what it seems. It is 'inverted': the wind has mined away the material around the channel: this ghost river now stands proud from the landscape.

Maps of Mars tell a story of change, perhaps even of the rise and fall of life. For Dr Devon Burr, of Northern Arizona University, one of the cartographers who worked on our map, they are stories we should learn from, for they show how climate change can radically alter worlds. 'Understanding the fascinating story of Mars's evolution' Dr Burr explains, may 'help people recognize the magnitude of possible change in planetary climates'.

One day the whole of Mars will be mapped in as much detail as Earth is. For now, we have just a few areas; bright patches of knowledge to add to older maps. One of the striking features of these older maps is ice caps. They were first spotted in 1672 by Dutch polymath Christiaan Huygens and they wax and wane with the seasons.

The map we reproduce here is of the northern cap and the ice is shown in white. The ice at the poles was once thought to be made up entirely of frozen carbon dioxide; puncturing dreams of melting it into water and so creating a habitable planet. However, in 2022 scientists used radar to discover that there may be huge qualities of water hidden underneath the Martian surface after all. Professor Neil Arnold, from Cambridge University, explains that 'radar data make it much more likely that at least one area of subglacial liquid water exists on Mars today'. He points to another profound implication: 'Mars must still be geothermally active in order to keep the water beneath the ice cap liquid.'

There's water, there is heat: visions of creating habitation on Mars are back on the table. But we should also be reminded of something else: Olympus Mons. Maybe not dead but dormant. The colonization of Mars may be possible. But settlement will not be carefree. Future residents of Mars will need to keep a weather eye on the grand old mountain that looms so large and so high in the distance.

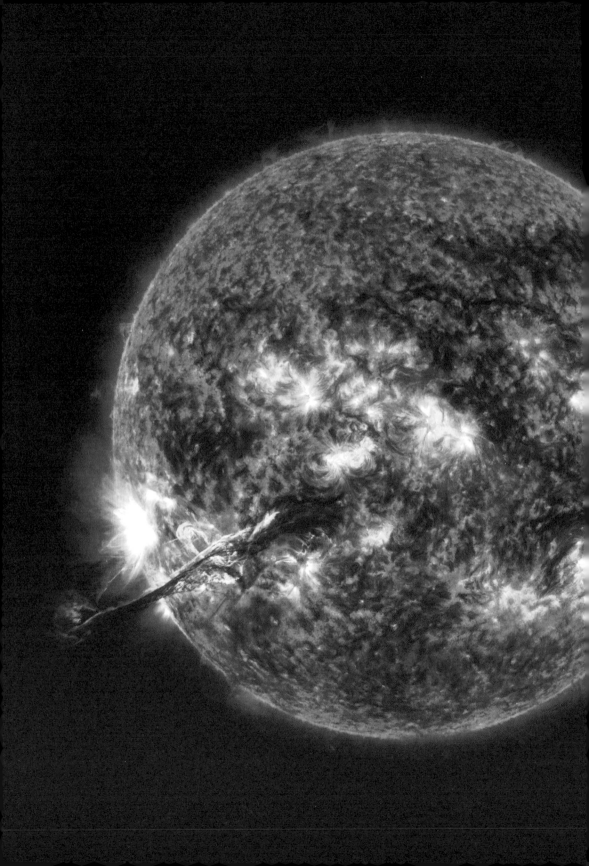

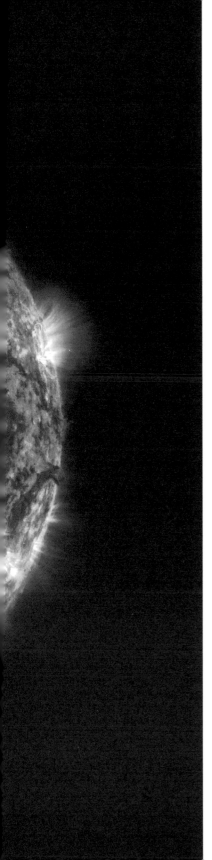

Map 38
The Sun, Showing Solar Flares and a Coronal Mass Ejection, 2012

A journey across the Sun

We're standing on the Sun and it's 5,700°C (10,292°F). In each direction we see a limitless golden plain, dense with ruptures and gashes, and on every horizon pillars of white light bolting into a silver sky. The verticals are dancing and this plateau is boiling. This is a real landscape. The Sun is not whisp, a vapour, a puff of flame: it has topography and weather, the stuff of geography.

Let's keep moving; let's go on a journey. There is much to explore and today promises some spectacular sights. Down in the south-west there are explosive events that have the power to shake worlds. If we never stopped to rest, it would take about one hundred years to walk all the way around the Sun so let's use our imagination; it's much quicker. If it's only fireworks we're after and we don't have long, then we should give the poles a miss. The north and south polar regions of the Sun are called the Quiet Sun for a reason. You won't find sunspots, mass ejections or leaping sky loops in the Quiet Sun.

What the Quiet Sun does offer are the best opportunities to see the less photographed side of our familiar star. We can witness its ordinary places, get to know its granules. It's an odd choice of word. 'Granules'. They are each about the size of France, often bigger, and there are about four million of them. It's these granules that give the Sun its mottled, speckly appearance. They are upwellings, convection cells that take plasma from the deeps, and deposit it on the surface, where it cools and falls back down though inky lanes that border each granule.

Away from the poles it does not take long before we start to see much bigger structures, arcs of matter that stream above the surface. Every day, every minute, on the landscape of the Sun is different. We're visiting on 31 August, 2012 and our map captures the surface of the Sun in one of its moments of everyday drama. A huge coronal mass ejection is underway to the bottom left of our map, travelling at 1,450 kilometres (about 900 miles) a second and blasting deep into space.

On Earth we are showered with the Sun's weather. These events create brilliant auroras across Earth's night-time skies. Such colourful auroras are the painted face of a remorseless foe. Without the protection of Earth's magnetic field, the Sun's weather would flay our planet, stripping out the atmosphere and putting an end to the possibility of life.

As a result of its bursts and jets, the Sun loses up to two million tonnes of mass every second. It makes little difference. For the Sun, two million tonnes is a mere

FIGURE XLVIII.

Imagined landscapes of the Sun, Allain Manesson Mallet, 1683 spoonful. The Sun contains 99.8 per cent of all the mass in the entire solar system (all the planets and moons make up a mere 0.2 per cent).

As we continue our tour, let's not forget to keep our ear plugs firmly wedged in. The Sun is loud. A continuous storm of nuclear explosions makes it the noisiest place in the solar system. The volume shakes everything to bits, including our bodies, which are being vibrated and pummelled into atoms. It's a sombre thought, so let's not dwell on it, especially now, as one of the most famous of the Sun's special places is coming into view.

Sunspots cluster round the wide central belt of the Sun. They are small tapestries of black, almost too small to make out on our map. What we see are the huge white and gold strands of energy that leap and weave in elliptical skeins all around them. The spots are calm, cooler patches surrounded by frenetic hot zones. They are also sites of intense magnetism, about 2,500 times stronger than the Earth's magnetic field.

Here is a wonderful example: seen at the centre of the map, to the right of the coronal mass ejection is a bright multi-centred landscape of loops and holes. It's complex territory, with several main sunspots that snake together and link to smaller ones. As we're so close we can peer down over the edge and into the spot. They are not mountains but great basins in the surface of Sun. It's hard to know how deep this one is; but if we fell in we'd be falling for a long time.

The bigger sunspots get named and tagged, each one given a number by the USA's National Oceanic and Atmospheric Administration, who are building a record of their comings and goings. Sunspots decay quickly, lasting at most a few months and they are far from static, scuttling at speed over the surface of the Sun. Flotillas of spots are carried hither and thither by the rotations of the star.

Humans are Sun watchers, always have been and always will be. There are histories in many cultures of its motions and cycles. Sunspots have been seen and recorded for thousands of years. The idea that the Sun has topography is not new. The notion that there is more to be seen here than celestial fire has long made instinctual sense.

The very first work of science fiction, which was written by the Syrian satirist Lucian of Samosata, in the second century CE, recounts a war between the inhabitants of the Sun and the Moon. There is a French print from 1683 (pictured opposite) with a detailed map of a kindly Sun, radiant and nurturing above a pastoral Earth. It reflects a fond hope that the Sun is a brilliant version of our own world; much like Earth but gilded and glorious. Today we know some things long suspected, that the Sun is neither kind nor cruel; that it is nothing like the Earth, and the closer we get to it, the more it sends our imagination reeling.

Overleaf: Auroras: The Sun's
weather hits the Earth

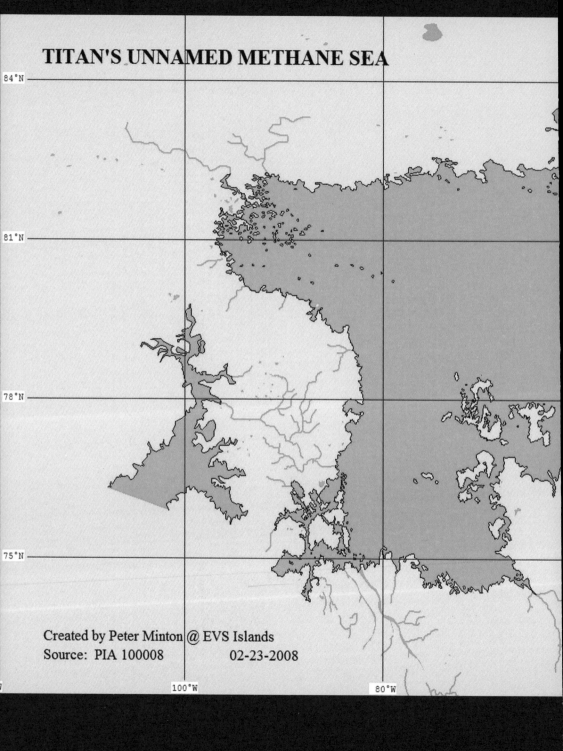

TITAN'S UNNAMED METHANE SEA

84°N

81°N

78°N

75°N

100°W

80°W

Created by Peter Minton @ EVS Islands
Source: PIA 100008   02-23-2008

0   100   200   300   400 km

1 cm =75.00 km

40°W

Map 39
**Ligeia Mare,
Titan, 2008**

Lakeside on Titan

Shall we take a walk, lakeside? Pretty creeks tumble through the rocks, and islands scatter the bay, large and small. Perhaps a rowboat might be pleasant? There is so much to explore and it's another windless day. We're on the shore of Ligeia Mare, a great lake one and half times the size of Lake Superior, and we're on Titan, the largest moon of Saturn.

The lake has a crenulated coast, some 2,000 kilometres (1,243 miles) long. The sculpting of landscapes by liquid is something we see everywhere on Earth but here on Titan it's just as common: there are coasts, valleys, estuaries and river mouths. It's all so familiar.

Yet before we push out our little boat let's recall that name, Ligeia Mare. *Mare* is Latin for sea but *Ligeia* carries a warning. Ligeia was a Greek siren, who lured sailors to their doom. Perhaps all is not what it seems.

Those gentle lapping waves are not water but liquid methane. If you dip your toe in it will instantly freeze and break off. The temperature of this lake is –180°C (–292°F). This is a sea of refrigerant, even down to its deepest reaches, some 200 metres (656 feet) below. Titan has a thick atmosphere, a golden shroud of nitrogen and methane and, along with its river and lakes, it has clouds and rainfall, a complete hydrological cycle. But none of this is water.

Recent images of Titan's surface indicate that some of Ligeia Mare's islands are not what they seem. A 260 square kilometre (100 square mile) island has been seen to rise and fall and been nicknamed Magic Island. Maybe it's made of ice, rising and melting. Maybe not. There are plenty of questions. For the time being, Ligeia Mare holds on to its secrets.

There are 150 moons in our Solar System. Titan is special. It is the most Earth-like planet we know. The desire to recognize it as a home-from-home is powerful. Another way to look at Titan's unhealthy chemical mix is as potential energy. Ligeia Mare is a massive lake of methane. What does methane do? It burns; it's fuel. And Ligeia Mare is just one small, local example. Titan has huge dunes of hydrocarbons. According to Ralph Lorenz, one of the scientists currently studying Titan, it is 'just covered in carbon-bearing material – it's a Mega factory of organic chemicals'. Titan has far more fuel to burn than Earth and it is for this reason that, according to some pundits, it is in pole position for human colonization.

Almost limitless power could be generated from those energy deposits. Abundant nitrogen in the atmosphere may also be useful, such as in making fertilizer (nitrogen is the main ingredient in many fertilizers). And deep beneath its icy shell Titan has a trump card. We do not find water on the surface but it is now thought that Titan has a huge sub-surface ocean of water several times larger than all the Earth's oceans.

Robert Zubrin, an aerospace engineer who has long argued for the colonization of Mars, is a big fan of Titan. 'In certain ways' he says, 'Titan is the most hospitable extra-terrestrial world within our solar system for human colonization'.

Any colonizer would feel the weight of Titan's thick atmosphere. All that wraparound cloud and gas means that atmospheric pressure is about sixty per cent heavier than on Earth. Yet it does not follow that life on Titan would be lived in the slow lane. Titan has high pressure but, being a small planet, it has also low gravity. This unique ratio of atmospheric density and low surface gravity means that getting around on Titan would be easier than on Earth. In his techno utopian treatise *Entering Space: Creating a Spacefaring Civilization*, Zubrin claims that that anyone could take a pair of wings and fly across Titan's surface.

On the website 'Humans to Titan' I read that 'A human in a hang glider could comfortably take off and cruise around powered by oversized swim–flipper boots – or even take off by flapping artificial wings. The power requirements are minimal – it would probably take no more effort than walking'.

Let's not order those 'swim–flipper boots' just yet. There is a problem. Titan is about 1.2 billion kilometres (¾ billion miles) away. It would take years to get there. If we do, people may fondly recall the first postcards sent from its surface, all thanks to Huygens, a small NASA probe dropped onto the surface in 2005. Our photograph (reproduced below) was taken during its descent, from a height of 16.2 kilometres (10 miles) and shows drainage channels twisting through hills on the way to a shoreline.

Hopefully it won't be too long before we get another look. One of the most exciting space missions planned for the next few decades will see a return to Titan. NASA's Dragonfly spacecraft is due to launch in 2027 and will arrive seven years later in 2034. It will position a robotic flight capable vehicle on Titan's surface, a rotor machine able to travel at about 36 kilometres (22 miles) per hour and fly up to an altitude of 4 kilometres (2½ miles). One of Dragonfly's aims is to investigate the potential for human colonization. It also wants to find out if the planet's hydrocarbons reveal any of the chemical building blocks of life.

Dragonfly is a new type of space mission, focused on astrobiology, the meeting point of the science of life and space exploration. What will it find? Hopefully, one morning, in 2034, we'll wake up to find out.

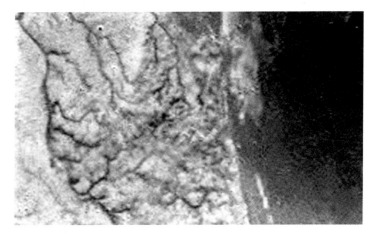

A photograph of the surface of Titan, showing drainage systems, 2005

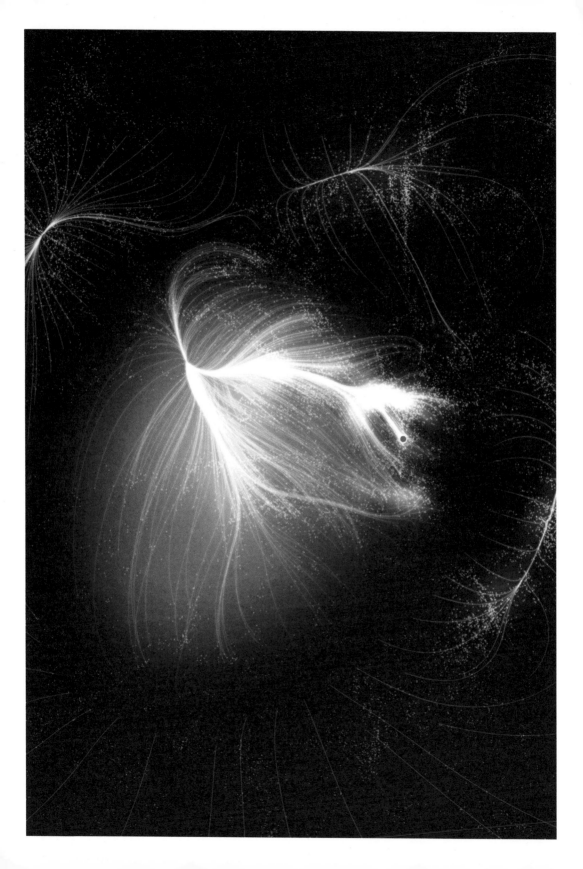

Map 40

Laniakea Supercluster: Our Galaxy, the Milky Way, is Under the Tiny Red Dot

Our new home: the Laniakea Supercluster

There is a new place we can call home. It has a Hawaiian name, Laniakea, which means 'immense heaven'. There is a small red dot on our map. Under it there are a lot of galaxies, including our own, the Milky Way. Take another look, take a pause, take a leap.

The map produced by the Hawaii team is a map of cosmic paths, the migratory routes of galaxies propelled by gravity. How big is it? To answer that we need to know that one light year is the distance light takes to travel in one year, which is 9.46 trillion kilometres (5.88 trillion miles). Laniakea is 520 million light years across.

We have known for a long time that our solar system is part of something bigger. Eventually it was discovered that our Sun and all its revolving planets are located amid many other solar systems, which together make up one filament, one tendril, of one arm of the hundred billion stars and billions more planets that constitute the spiralling cosmos we call the Milky Way.

Astronomers knew about galaxies beyond our Milky Way. It was thought they were disconnected colossi, hurtling randomly through space away from the Big Bang. This was a mistake. Galaxies are not the biggest thing in the universe. They are part of larger structures. Structures like Laniakea. To know our place in the universe we need to know their names and understand our relationship to them.

There are galaxies, then there are galaxy clusters and, bigger still, superclusters. Laniakea is a supercluster. It is made up of between 100,000 and 150,000 galaxies. It's our supercluster, it's where we live and our portion of it, the bright rivulet where that red dot is, is called the Virgo cluster.

The universe has structure. Its two great forces, expansion and gravity, braid and sculpt the galaxies into rivers of motion and matter. The resemblance to rivers, to channels of water, is more than poetic: the scientists who developed this map talk of the superclusters as forming 'watersheds', with galaxies falling one way or the other, into one cluster or another; like rainfall draining into one valley or another, getting drawn by gravity, and tumbling seaward.

It's a pattern found at many scales. On any beach, such as the one pictured here, you can find Laniakea, or something like it, and look at its branching flows. Just tiny rivulets, snouting across sand grains. It's there in many places, in the pattern of veins on my wrist; in a leaf or a branching tree. A pattern of flows is echoed across the universe, making the shape of Laniakea appear as familiar, natural and beautiful.

These threaded and twisting motions are the stuff of maps and it is to maps that space scientists are

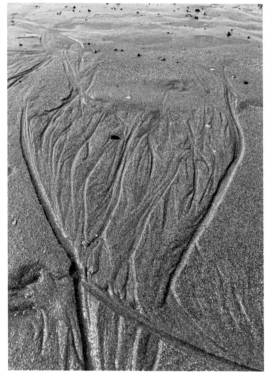

Laniakea on the beach

turning, not simply to illustrate the cosmos, but to understand it. Like others in the team at the University of Hawaii that modelled Laniakea, Hélène Courtois calls herself a 'cosmo-grapher', a mapper of the large-scale features of the observable universe.

Laniakea is made up of different zones of high density. Apart from Virgo, which is our bit, it is important to familiarize ourselves with the central branching Y-shape that dominates this map. This is the Hydra–Centaurus Supercluster but it has another, more telling, name. the 'Great Attractor'. It is towards this that the other parts of Laniakea are drawn, gravitationally pulled and sped. Laniakea is a river, a 'watershed', but it is folding in on itself, draining itself into the sink hole of the Great Attractor.

The map of Laniakea helps us to understand why the Milky Way is travelling at 600 kilometres (about 373 miles) per second: it's being pulled along by other galaxies in our supercluster, and they are all draining towards the Great Attractor.

At the same time as gravity is folding and collapsing Laniakea, the forces of expansion, which keep pushing the universe outwards, are dispersing it. This new home, this 'thing' of which we are a part, is neither stable nor bound together but temporary. The Great Attractor is not great enough and one day Laniakea will break up, and our galaxy will spin free of its cousins. So too our solar system, and every last molecule on Earth. The Hawaiian team is perfectly clear. Things fall apart: the centre cannot hold. Laniakea is a river, and it has structure, but these 'infalling motions on large scales are only perturbations', they write, all 'galaxies except those in the immediate vicinity of clusters and groups are flying apart'.

Another layer of complexity can also be added: Laniakea is itself being attracted toward an even larger supercluster, called the Shapley supercluster. Shapley is made up of 50 million billion solar masses and is located about 500 million light years away from Earth.

Does nothing hold? Must everything, me and Laniakea both, be swallowed up, scattered, obliterated? As an ordinary Laniakean, briefly content in my portion of the universe, I'm annoyed: just when I was getting comfortable I'm told, forget all that, in the greater scheme of things you're nothing. Laniakea is dust. It's what we are always told and it's always right: every channel and trickle is guided by a larger, more powerful, flow.

References

A 9,000-year-old town map
James Mellaart, 'Excavations at Çatal Hüyük, 1963, Third
    Preliminary Report', Anatolian Studies, 14, 1964.
Ali Umut Türkcan, cited in 'New Neighborhood Found in
    Çatalhöyük', Hürriyet Daily News, July 15, 2021.

The first modern map
Yu the Great, cited by J.G. Cheock, China: Myth
    or History?, 2017.
Joseph Needham, Science and Civilisation in China:
    Volume 3: Mathematics and the Sciences of the
    Heavens and the Earth, 1959.

The oldest map of the British Isles
William Camden, 'Essex', A Vision of Britain Through Time',
    visionofbritain.org.uk
Jim Crumley, The Great Wood: The Ancient Forest Wood,
    2014.

When China discovered the world?
The Economist, 'China beat Columbus to it, perhaps',
    Jan 12, 2006.
Translations and citations from 1418 map: Mo Yi Tong,
    'Chinese 1418 Map', undated, available atmyoldmaps.com/
    late-medieval-maps-1300/2363-1418.pdf
Philip Snow, The Star Raft: China's Encounter with Africa,
    1988.
Geoff Wade, 'The "Liu/Menzies" World Map: A Critique',
    e-Perimetron, 2, 4, 2007.

**The world on its head:.The Ottoman Empire
    eyes the New World**
Tarih-i Hind-i Garbî citations by Emir Karakaya, 'Tarih-i Hind-i
    Garbî as a Manifestation and as a Mirror', 2016, available
    atacademia.edu
See also: Thomas D. Goodrich, The Ottoman Turks and
    the New World: A Study of Tarih-i Hind-i Garbi and
    Sixteenth-century Ottoman Americana, 1990.

Lord Quetzalecatzin's map: Aztecs versus colonial power
Barbara E. Mundy, 'Mesoamerican Cartography', in David
    Woodward and G. Malcolm Lewis (Eds) The History of
    Cartography, 2, 3: Cartography in the Traditional African,
    American, Arctic, Australian, and Pacific Societies, 1998
For references to descriptions of indigenous maps from
    Hernan Cortes and Bernal Diaz del Castillo see: Mundy,
    'Mesoamerican Cartography'.

All the world's people, as seen from Japan
Citations from map: Elke Papelitzky, 'A Description and
    Analysis of the Japanese World Map Bankoku sōzu in
    Its Version of 1671 and Some Thoughts on the Sources
    of the Original Bankoku sōzu', Journal of Asian History
    48, 1, 2014.
See also: Kazuta Kaunno 'Cartography in Japan', in J.
    B. Harley and David Woodward (Eds), The History of
    Cartography, 2, 2: Cartography in the Traditional East
    and Southeast Asian Societies, 1994.

A sailor's map made of sticks and shells
Ben Finney, 'Nautical Cartography and Traditional Navigation
    in Oceania', in David Woodward and G. Malcolm Lewis
    (Eds) The History of Cartography, 2, 3: Cartography in
    the Traditional African, American, Arctic, Australian, and
    Pacific Societies, 1998
See also: Cynthia Smith, 'The Unique Seafaring Charts of the
    Marshall Islands', Library of Congress, 2021, blogs.loc.
    gov/maps/2021/11/the-unique-seafaring-charts-of-the-
    marshall-islands/

A shattered tear drop: a journey to another shore
Richard Carnac Temple, The Thirty-Seven Nats: A Phase
    of Spirit Worship Prevailing in Burma, 1906.
See also: Frank Jacobs, 'Our Alien World: A Burmese
    Spirit Map', December 3, 2015, Big Think, bigthink.com/
    strange-maps/a-burmese-map-of-the-world/

If Africa had not been colonized
Nikolaj Jesper Cyon, 'ALKEBU-LAN 1260 AH',
    cyon.se/#/alkebulan-1260-ah/

An empire made visible

1863 imperial decree, cited by Alastair Bonnett, 'Russia's colonial legacy and the war in Ukraine', Geographical, March 16, 2022, geographical.co.uk/news/russias-colonial-legacy-and-the-war-in-ukraine.

See also: G. Patrick March, Eastern Destiny: Russia in Asia and the North Pacific, 1996.

The German USA

For further reading see: Sudie Doggett Wike, German Footprints in America: Four Centuries of Immigration and Cultural Influence, 2022.

American Community Survey, 2022, see: United States Census, 'People Reporting Ancestry', data.census.gov/table/ACSDT1Y2022.B04006?q=ancestery

Who owns Alaska?

Byron Mallott cited in PBS, 'Native Land Claims in Alaska', undated, PBS, pbs.org/wgbh/americanexperience/features/pipeline-native-land-claims-alaska/#:~:text=The%20state%20and%20federal%20governments,to%20remove%20this%20potential%20roadblock.%22

Patricia Cochran, cited by Mary Robinson, Climate Justice: Hope, Resilience, and the Fight for a Sustainable Future, 2018

Alaska Federation of Natives convention: see Alex DeMarban 'AFN Declares 'State of Emergency' for Climate Change', Anchorage Daily News, October 20, 2019.

A world-leading environmental plan from Africa

Jim Morrison, 'The "Great Green Wall" Didn't Stop Desertification, but it Evolved into Something That Might', Smithsonian Magazine, August 23, 2016.

Nigerian Government, 'GGW Nigeria, NAGGW Signs MoU with Groasis B.V And Boplas Industry Limited to Establish Waterboxx Production Line in Nigeria', National Agency for the Great Green Wall, April 19, 2023, ggwnigeria.gov.ng/naggw-signs-mou-with-groasis-b-v-and-boplas-industry-limited-to-establish-waterboxx-production-line-in-nigeria/

New maps for a new superpower

Hao Xiaoguang, cited by China News Agency, 'Looking at the World Horizontally and Vertically, What's the Big Difference? Interview with Hao Xiaoguang, the Compiler of the Vertical World Map', tellerreport.com, April 14, 2021, tellerreport.com/life/2021-04-14-looking-at-the-world-horizontally-and-vertically--what-s-the-big-difference--%E2%80%94%E2%80%94interview-with-hao-xiaoguang--the-compiler-of-the-vertical-world-map.rJVIEj2E8d.html

Manny Mori, cited by Harry Kazianis, 'China Tricked a Yale Conference To Believe It Claimed the Entire Pacific Ocean', The National Interest, March 14, 2021, nationalinterest.org/blog/reboot/china-tricked-yale-conference-believe-it-claimed-entire-pacific-ocean-180202

Artic ice cover loss, see: National Snow and Ice Data Center, 'Arctic, low. Antarctic, whoa', July, 2023, nsidc.org/arcticseaicenews/2023/07/

All heading south! Antarctica is getting busy

John Davis, cited by David Day, Antarctica, 2019.

Arlie McCarthy, 'Antarctica's unique ecosystem is threatened by invasive species "hitchhiking' on ships", The Conversation, January 11, 2022.

See also: McCarthy, A.H., Peck, L.S., & Aldridge, D.C. 'Ship traffic connects Antarctica's fragile coasts to worldwide ecosystems' PNAS, January 2022.

David Aldridge, cited by St Catherine's College, 'Invasive Species "Hitchhiking" on Ships Threaten Antarctica's Ecosystems, Tuesday 11 January 11, 2022, caths.cam.ac.uk/hitchhiking

Leyla Cardenas, cited in La Prensa Latina, 'Study: Mussels Could Invade Antarctica, Severely Threaten its Biodiversity', March 31, 2020, La Prensa Latina, laprensalatina.com/study-mussels-could-invade-antarctica-severely-threaten-its-biodiversity/#:~:text=A%20mussel%20settlement%20had%20never,and%20rapidly%20dominate%20their%20surroundings.%E2%80%9D

China's high-speed revolution
For further reading see: Martha Lawrence, Richard
    Bullock, and Ziming Liu, China's High-Speed Rail
    Development, 2019.

A transparent map of a three-dimensional city
Tomoyuki Tanaka, cited in 'Tomoyuki Tanaka – Japan's
    Major Landmarks Drawn as if Seen by X-ray' Wepresent,
    September 14, 2016, wepresent.wetransfer.com/stories/
    tomoyuki-tanaka-on-x-ray-vision-and-shinjuku-station

Big maps for big data
Barrett Lyon, cited by Jack Linshi, 'See What the Internet
    Actually Looks Like', Time, July 13, 2015, time.
    com/3952373/internet-opte-project/
and by Lily Hay Newman, 'A Trippy Visualization Charts the
    Internet's Growth Since 1997', Wired, February 21, 2021,
    wired.com/story/opte-internet-map-visualization

Google's chameleon maps
For further reading see: Greg Bensinger, 'Google Redraws
    the Borders on Maps Depending on Who's Looking',
    Washington Post, February 14, 2020.

A women-friendly city is something to be
Andrea Gorrini, Dante Presicce, Rawad Choubassi, and Ipek
    Nese Sener, 'Assessing the Level of Walkability for Women
    using GIS and Location-based Open Data: The Case of
    New York City', Findings, 2021.

Overwhelmed by the noise of cars and trucks
Fausto Rodriguez-Manzo, et al. 'Towards an Acoustic
    Categorization of Urban Areas in Mexico City',
    INTER-NOISE and NOISE-CON Congress and
    Conference Proceedings, 253, 1, 2016.
World Bank air pollution statistic, see: Copenhagen
    Consensus, 'Mexico Perspective: Air Pollution',
    2023, copenhagenconsensus.com/publication/
    mexico-perspective-air-pollution
Bruitparif, cited by Giovanna Coi and Aitor Hernández-
    Morales, 'Europe Struggles to Turn Down Volume on
    Deadly Traffic Noise', Politico, September 7, 2022, politico.
    eu/article/eu-france-paris-european-green-deal-noise-
    pollution-in-cities-sounds-like-a-problem/
See also: bruitparif.fr/
University of Michigan study: Richard Neitzel, et al. 'Exposures
    to Transit and Other Sources of Noise among New York
    City Residents', Environmental Science and Technology,
    46, 1, 2012.
UK Health Security Agency, 'Noise Pollution: Mapping
    the Health Impacts of Transportation Noise in England',
    June 29, 2023, ukhsa.blog.gov.uk/2023/06/29/
    noise-pollution-mapping-the-health-impacts-of-
    transportation-noise-in-england/
European Environment Agency, 'Noise Pollution is a Major
    Problem, Both for Human Health and the Environment',
    May 11, 2021, eea.europa.eu/articles/noise-pollution-is-
    a-major#:~:text=Looking%20at%20the%20current%20
    data,suffer%20chronic%20high%20sleep%20
    disturbance.

Mapping the smells of a seaside vacation
Kate McLean, citations from website: sensorymaps.com/
Kate McLean, cited by Frank Jacobs, 'A Map of the Smells
    of Newport, Rhode Island', Big Think, January 8, 2014,
    bigthink.com/strange-maps/638-nil-the-nose-knows-
    an-olfactory-map-of-newport-ri/
Kyiv smellscape, see also: Kate McLean, 'Polyrhythmia of
    the smellwalk: Mapping multi-scalar temporalities',
    undated, available at: repository.canterbury.ac.uk/
    download/83c25371016140a060b884392456ea78b7d-
    3f9a8c7e3949df635ce2fe98a6d84/16313311/Mapping_
    Smellwalking_Mclean_EDI_2017_edited.pdf
University of California study: Jess Porter, et al.,
    'Mechanisms of Scent-tracking in Humans',
    Nature Neuroscience, 10, 1, 2007.

A game of love

Madeleine De Scudéry, Les femmes illustres, ou les harangues heroïques de Monsieur de Scudéry, 1642.

Madeleine De Scudéry, Clélie, an Excellent Romance: The Whole Work in Five Parts, 1678.

See also: Madeleine De Scudéry, The Story of Sapho, 2003.

Wandering ghosts: the art of GPS

Jeremy Wood, cited in 'Interview with Jeremy Wood', GPS Maps, gpsdrawing.com/maps.html and 'Interview with Jeremy Wood', LEA Interview, 2011, chrome-extension:// efaidnbmnnnibpcajpcglclefindmkaj/ leoalmanac.org/ wp-content/uploads/2011/07/wood.pdf

W. G. Sebald, The Rings of Saturn, 1998.

Iain Sinclair, London Orbital, 2002.

Mapping the human brain, one sliver at a time

Estimates on computer memory space, time needed for counting connections manually and quotes from Alex Shapson-Coe, see: Jason Dorrier, 'Google and Harvard Unveil the Largest High-Resolution Map of the Brain Yet', Singularity Hub, June 6, 2021, singularityhub. com/2021/06/06/google-and-harvard-unveil-the-largest- high-resolution-map-of-the-brain-yet/

Jeff Lichtman, cited by Tom Metcalfe, 'This is Your Brain, in Glorious Color', NBC, June 15, 2021, nbcnews.com/science/ science-news/brain-glorious-color-rcna1192

See also: Neuroglancer, microns-explorer.org/visualization

Dancing geography: how bees draw maps

Fernando Wario et al., 'Automatic Detection and Decoding of Honey Bee Waggle Dances', PLoS ONE 12, 13, 2017.

Matthew Hasenjager, cited in 'Bees Prioritise their Unique Waggle Dance to Find Flowers', Royal Holloway, February 7, 2020, royalholloway.ac.uk/research-and-teaching/ departments-and-schools/biological-sciences/ news/bees-prioritise-their-unique-waggle-dance- to-find-flowers/

Karl von Frisch, The Dance Language and Orientation of Bees, 1967. (Translation of Tanzsprache und Orientierung der Bienen, 1965.)

Wood wide web: how trees and fungi help each other out

Kevin Beiler, et al., 'Architecture of the Wood-wide Web: Rhizopogon Spp. Genets Link Multiple Douglas- fir Cohorts', New Phytologist, 185, 2010.

Dave Hansford, 'The Wood Wide Web', New Zealand Geographic, 2017, nzgeo.com/stories/the-wood- wide-web/

Yuan Yuan Song, et al., 'Defoliation of Interior Douglas-fir Elicits Carbon Transfer and Stress Signalling to Ponderosa Pine Neighbors through Ectomycorrhizal Networks', Sci Rep, 5, 8495, 2015.

Suzanne Simard and marigold study cited by Dave Hansford, 'The Wood Wide Web'.

Earth power: how earthquakes reveal a hidden planet

Jeroen Tromp, cited by Catherine Zandonella, 'Frontier Beneath our Feet: Seismic Study Aims to Map Earth's Interior in 3-D', March 12, 2015, Princeton University, princeton.edu/news/2015/03/12/frontier-beneath-our- feet-seismic-study-aims-map-earths-interior-3-d

Planet potato: a map of gravity

Eric Hinds, 'A New Year, a New geoid - OSTN15', Landform Surveys, January 23, 2023, landform-surveys.co.uk/ news/general-updates/new-year-new-geoid- ostn15/#:~:text=Calf%20Hill%20is%20located%20 in,Hill%20now%20measures%20at%20609.606.

Vertical extremes: from Challenger Deep to Tilicho Lake

John Murray, cited in 'The History of the Challenger Expedition', Challenger Society for Marine Science, challenger-society.org.uk/History_of_the_challenger_ expedition

Stephen Stukins, cited by James Ashworth, 'HMS Challenger: How a 150-year-old Expedition Still Influences Scientific Discoveries Today', September 6, 2022, Natural History Museum, nhm.ac.uk/discover/news/2022/september/ hms-challenger-how-150-year-old-expedition-still- influences-scientific-discoveries-today.html

On Lake Baikal see: Tibi Puiu, 'Exploring Baikal: The World's Deepest and Oldest Lake', August 26, 2022, ZME Science, zmescience.com/feature-post/natural-sciences/ geography/exploring-baikal-the-worlds-deepest- and-oldest-lake/

There is only one ocean

General Bathymetric Chart of the Oceans website, gebco.net/about_us/

200 years to map oceans, *see* General Bathymetric Chart of the Oceans, 'Frequently Asked Questions', gebco.net/about_us/faq/

Proposal form for naming underwater features, *see* 'Standardization of Undersea Feature Names (Guidelines, Proposal Form Terminology)', link at General Bathymetric Chart of the Oceans, iho.int/en/bathymetric-publications

New Zealand bathymetry, *see* National Institute of Water and Atmospheric Research, 'New Zealand Bathymetry', niwa.co.nz/our-science/oceans/bathymetry/further-information

Oceans and climate, *see* Francisco de Melo Viríssimo and Elizabeth Robinson, 'What Role do the Oceans Play in Regulating the Climate and Supporting Life on Earth?', February 28, 2023, LSE, lse.ac.uk/granthaminstitute/explainers/what-role-do-the-oceans-play-in-regulating-the-climate-and-supporting-life-on-earth/

Biomass data, *see* Hannah Ritchie, 'Oceans, Land and Deep Subsurface: How is Life Distributed Across Environments?', April 26, 2019, Our World in Data, ourworldindata.org/life-by-environment

Sundaland: a hidden ancient homeland, now under the sea

Steven Mithen, After the Ice, A Global Human History, 20,000 – 5000 BC, 2004.

Flints and antler hoards, *see* Jim Leary, The Remembered Land: Surviving Sea-level Rise after the Last Ice Age, 2015

Sundaland as Atlantis, cited in 'Sundaland', Antlantipedia, June 11, 2010, atlantipedia.ie/samples/sundaland/

Danny Hilman Natawidjaja, Plato Never Lied, Atlantis is in Indonesia, 2013.

Dhani Irwanto, Atlantis: The Lost City is in Java Sea, 2015.

Sundaland as Eden, *see* Stephen Oppenheimer, Eden In The East: Drowned Continent of Southeast Asia, 1998.

Sundaland pyramids, *see* Robert M. Schoch and Robert Aquinas McNally, Voyages of the Pyramid Builders: The True Origins of the Pyramids from Lost Egypt to Ancient America, 2003.

Europe in a world without ice

Christopher Bretz, citations from website, christopherbretz.ca/pf/sealevelrise/

Sea-level, *see* Christina Nunez, 'Sea Levels are Rising at an Extraordinary Pace', April 10, 2023, National Geographic, nationalgeographic.com/environment/article/sea-level-rise-1

Alastair Bonnett, The Age of Islands, 2020.

Nile delta, *see* Saline Agriculture, 'Salinity Problems in Egypt', January 25, 2022 saline-agriculture.com/en/news/salinity-problems-in-egypt#:~:text=Causes%20of%20soil%20salinity,be%20classified%20as%20salt%2Daffected.

Michael Bevis cited by Oliver Milman, 'Greenland's Ice Melting Faster than Scientists Previously Thought', January 22, 2019, The Guardian.

Cooking Iberia: a map of extreme heat

300 years to create one centimetre of fertile soil study, cited by Roger Harrabin, 'Climate change being fuelled by soil damage – report', April 29, 2019, BBC, bbc.co.uk/news/science-environment-48043134

Valencia residents cited by AFP, 'Spanish City Shatters Heat Record', August 10, 2023, France 24, france24.com/en/live-news/20230810-spanish-city-shatters-heat-record

Home Ignition Zones, *see* 'Defensible Space', Tri-County Firesafe Working Group, 2023, tcfswg.org/be-prepared/defensible-space/ and 'Make your house safe with Defensible Space', Hills Conservation Network, hillsconservationnetwork.org/make-your-home-ignition-resistant

Great Plains Shelterbelt *see*, Matthias Gafni, 'An 800-Mile Firebreak Once Protected California's Forests From Flames. What Happened?', November 19, 2020, San Francisco Chronicle, sfchronicle.com/california-wildfires/article/An-800-mile-firebreak-once-protected-15713546.php

The hidden great lake

Mathieu Morlighem, et al., 'BedMachine v3: Complete Bed Topography and Ocean Bathymetry Mapping of Greenland from Multibeam Echo Sounding Combined with Mass Conservation', Geophysical Research Letters, 44, 2017.

Paul Bierman and Tammy Rittenour, 'When Greenland was Green: Ancient Soil from Beneath a Mile of Ice Offers Warnings for the Future', July 20, 2023, The Conversation, theconversation.com/when-greenland-was-green-ancient-soil-from-beneath-a-mile-of-ice-offers-warnings-for-the-future-209018

Kristian Hvidtfeldt Nielsen and Henry Nielsen, Camp Century: The Untold Story of America's Secret Arctic Military Base Under the Greenland Ice, 2021.

Emma Smith, cited by Jonathan Amos, 'Denman Glacier: Deepest Point on Land Found in Antarctica', December 12, 2019, BBC, bbc.co.uk/news/science-environment-50753113

Olympus Mons and new maps of Mars

USGS Astrogeology Science Center in Flagstaff, Arizona, see 'What Three New USGS Maps Reveal About Mars', January 12, 2023, USGS, usgs.gov/special-topics/planetary-geologic-mapping/news/a-martian-mons-mystery-paleo-climate-change-and

Map of Aeolis Dorsa, see 'What Three New USGS Maps Reveal About Mars'.

Devon Burr, cited in 'What Three New USGS Maps Reveal About Mars'.

Neil Arnold, cited by Sarah Collins, 'New Evidence for Liquid Water Beneath the South Polar Ice Cap of Mars, September, 29, 2022, Cambridge University, cam.ac.uk/stories/liquid-water-mars

A journey across the Sun

Sunspot classification see 'Sunspots/Solar Cycle', NOAA, swpc.noaa.gov/phenomena/sunspotssolar-cycle

On Lucian of Samosata, see Aaron Parrett, 'Lucian's Trips to the Moon', June 26, 2013, The Public Domain Review, publicdomainreview.org/essay/lucians-trips-to-the-moon/

Lakeside on Titan

Ralph Lorenz, cited in ESA, 'Titan's Surface Organics Surpass Oil Reserves on Earth', February 13, 2008, European Space Agency, esa.int/Science_Exploration/Space_Science/Cassini-Huygens/Titan_s_surface_organics_surpass_oil_reserves_on_Earth

Titan's subsurface water, see 'Titan', 2023, NASA science. nasa.gov/saturn/moons/titan/

Robert Zubrin, cited by Brian Wang, 'Best Colonization Target in Outer Solar System is Titan', March 14, 2015, nextbigfuture, nextbigfuture.com/2015/03/best-colonization-target-in-outer-solar.html

Robert Zubrin, Entering Space: Creating a Spacefaring Civilization, 1999.

Humans to Titan: humans-to-titan.org/

Dragonfly, see science.nasa.gov/mission/dragonfly

Our new home: the Laniakea Supercluster

R. Brent Tully, et al., 'The Laniakea Supercluster of Galaxies', Nature, 513, 2014, nature.com/articles/nature13674

Hawaiian team citations, from R. Brent Tully, et al., 'The Laniakea Supercluster of Galaxies'.

Hélène Courtois, Finding our Place in the Universe: How We Discovered Laniakea: The Milky Way's Home, 2019.

Index

Picture Credits

All Canada Photos/Alamy 170; Antiqua Print Gallery/Alamy 168; Arlie McCarthy, Marine ecologist, Helmholtz Institute for Functional Marine Biodiversity (HIFMB). Originally published in: Arlie H. McCarthy (Cambridge University and British Antarctic Survey) Lloyd S. Peck (British Antarctic Survey) and David C. Aldridge (Cambridge University), 'Ship traffic connects Antarctica's fragile coasts to worldwide ecosystems', PNAS, 2022: https://www.pnas.org/doi/epdf/10.1073/pnas.2110303118 74–5; Barrett Lyon/The Opte Project 85–6; Beiler, K.J., Durall, D.M., Simard, S.W., Maxwell, S.A. and Kretzer, A.M. (2010), Architecture of the wood-wide web: Rhizopogon spp. genets link multiple Douglas-fir cohorts. New Phytologist, 185: 543–553. https://doi.org/10.1111/j.1469-8137.2009.03069.x. © The Authors (2009). Journal compilation © New Phytologist (2009) 124; Berger, Shapson-Coe, Januszewski, Jain, and Lichtman (Harvard and Google) 116; Bill Waterson/Alamy 44–5; CESAR MANSO/AFP/Contributor/Getty 156; Chalermkiat Seedokmai/Getty 141; Chittka L: Dances as Windows into Insect Perception. PLoS Biol 2/7/2004: e216. https://dx.doi.org/10.1371/journal.pbio.0020216, Attribution 2.5 Generic (CC BY 2.5) 122; Christopher Bretz 150; Chronicle/Alamy 111; Commission Air/Alamy Stock Photo 115; David Pugmire, Oak Ridge National Laboratory, Ebru Bozdag, Colorado School of Mines, Jeroen Tromp, Princeton University 128; Dhani Irwanto, Sundaland: Tracing the Crade of Civilizations, Indonesia Hydro Media, 2019 146; Dimple Patel/Alamy 130; The original uploader was Engwar at English Wikipedia. Later versions were uploaded by Toytoy at English Wikipedia. - Transferred from en.wikipedia to Commons., Public Domain, https://commons. wikimedia.org/w/index.php?curid=1767473 22–3; ©ESA/AOES Medialab 136; ©ESA/HPF/DLR 134; ESA/ NASA/University of Arizona/KRT/Abaca Press/Alamy 175; European Union, Copernicus Sentinel-3 imagery 154–5; First Noise Map for Mexico City Metropolitan Area. Authors, Fausto Rodríguez-Manzo & the Laboratorio de Análisis y Diseño Acústico team at Universidad Autónoma Metropolitana-Azcapotzalco in collaboration with the Secretaria del Medio Ambiente del Gobierno del Distrito Federal. 2009–2011 98–9; filmstudio/Getty 97; François Chauveau - old engraving (see fr:Carte_de_Tendre), Public Domain, https://commons.wikimedia.org/w/index.php?curid=2616672 106; GENERAL BATHYMETRIC CHART OF THE OCEANS (GEBCO) WORLD OCEAN BATHYMETRY 142;

General Research Division, The New York Public Library. (1906). Hell according to the Burmese. Retrieved from https://digitalcollections.nypl.org/items/510d47d9-a8a7-a3d9-e040-e00a18064a99 49; Gobierno CDMX - HJ2A4913, CC0, https://commons.wikimedia.org/w/index.php?curid=115335826 101; Gorrini, A., Presicce, D., Choubassi, R., Sener, I.N. (2021). Assessing the Level of Walkability for Women Using GIS and Location-based Open Data: The Case of New York City. Findings. https://doi.org/10.32866/001c.30794. (Scientific Figure on ResearchGate. Available from: https://www.researchgate.net/figure/Results-of-the-proposed-Walkability-for-Women-Index-WWI-of-the-census-block-groups-and_fig2_357230457 [accessed 5 Oct, 2023], available via Creative Commons Attribution-ShareAlike 4.0 International) 94–5; © Greenland National Museum and Archives 43; History of America/Alamy 110; The Battle of Adwa 1896, 1972.250 © Horniman Museum. Photo: Patrick Marks 52–3; Howchou – Own work, CC BY–SA 3.0, https://commons.wikimedia.org/w/index.php?curid=59380944 78–9; Images & Stories/Alamy 8–9; Jacob d'Angelo after Claudius Ptolemaeus[1], Public Domain, https://commons.wikimedia.org/w/index.php?curid=3423762 20–1;

Jeremy Wood 112–3; Original research, design and artwork by Dr Kate McLean (sensorymaps.com) 102, 104; Krzysztof Baranowski/Getty 81; Lewis Hine/GRANGER - Historical Picture Archive/Alamy 61; (1593) The Codex Quetzalecatzin. [Mexico: Producer not identified] [Map] Retrieved from the Library of Congress, https://www.loc.gov/item/2017590521/ 30–1 and 33; Library of Congress, Prints & Photographs Division, photograph by Carol M. Highsmith [reproduction number, LC-DIG-highsm-25691], Fort Ross, a former Russian establishment on the west coast of North America in what is now Sonoma County, California. United States California Fort Ross, 2013. June. https://www.loc.gov/item/2013635136/ 57; Lin, M. (2016, January 08). Yu the Great. World History Encyclopedia. Retrieved from https://www.

Acknowledgements

Many thanks to everyone at Quarto who worked so hard to help me bring this book to publication, especially my editors Richard Green and Katerina Menhennet. Thanks also to the cartographers and scientists who have given their permission for us to use their creations, to Jeroen Tromp and Kevin Beiler, who provided helpful advice on their maps, copy-editor Lesley Malkin and my brother, Paul Bonnett, for his insights.

About the Author

Alastair Bonnett is Professor of Geography at the University of Newcastle. His travel and academic books have been translated into 19 languages and include *Off the Map*, *Beyond the Map*, *The Age of Islands*, *Multiracism*, and *How to be Original*.